T0220439

Second Edition

A FIRST COURSE
IN
INTEGRAL EQUATIONS

Second Edition

A FIRST COURSE
IN
INTEGRAL EQUATIONS

Abdul-Majid Wazwaz

Saint Xavier University, USA

World Scientific

NEW JERSEY • LONDON • SINGAPORE • BEIJING • SHANGHAI • HONG KONG • TAIPEI • CHENNAI

Published by

World Scientific Publishing Co. Pte. Ltd.
5 Toh Tuck Link, Singapore 596224
USA office: 27 Warren Street, Suite 401-402, Hackensack, NJ 07601
UK office: 57 Shelton Street, Covent Garden, London WC2H 9HE

Library of Congress Cataloging-in-Publication Data
Wazwaz, Abdul-Majid.
    A first course in integral equations / by Abdul-Majid Wazwaz (Saint Xavier University, USA). --
Second edition.
        pages cm
    Includes bibliographical references and index.
    ISBN 978-9814675116 (hardcover : alk. paper) -- ISBN 978-9814675123 (pbk. : alk. paper)
    1. Integral equations. I. Title.
    QA431.W36 2015
    515'.45--dc23
                              2015008111

British Library Cataloguing-in-Publication Data
A catalogue record for this book is available from the British Library.

Copyright © 2015 by World Scientific Publishing Co. Pte. Ltd.

*All rights reserved. This book, or parts thereof, may not be reproduced in any form or by any means,
electronic or mechanical, including photocopying, recording or any information storage and retrieval
system now known or to be invented, without written permission from the publisher.*

For photocopying of material in this volume, please pay a copying fee through the Copyright Clearance
Center, Inc., 222 Rosewood Drive, Danvers, MA 01923, USA. In this case permission to photocopy
is not required from the publisher.

Printed in Singapore

THIS BOOK IS DEDICATED TO

My wife, our son, and our three daughters
for supporting me in all my endeavors

Contents

Preface to the Second Edition

Within the last eighteen years since the publication of the first edition of *A First Course in Integral Equations*, the growth in the field of Integral Equations has been flourishing with many advances. The new developments, which complement the traditional concepts, present clear expositions of the main concepts and keys of Integral Equations. Moreover, a further significant recognition of the use of Integral Equations in scientific fields, engineering, and mathematics has developed. This recognition has been followed up with further achievements in research. Like the first edition, the second edition is helpful to a wide range of advanced undergraduate and graduate students in varying fields, as well as researchers in science, mathematics, and engineering. Some of the strengths of the new edition are the detailed treatments, clarifications, explanations of the new developments, discussions of the wide variety of examples, and the well-presented illustrations to aid the learner to better understand the concepts.

In editing this new edition, the following distinguishing features, above the pedagogical aims of the first edition, were highly considered:

1. Many new and remarkable developments have been added. The scope of each chapter is extended to contain these fascinating new findings.
2. The linear and the nonlinear integral equations were handled in a systematic manner with more methods and applications. This edition provides very systematic and detailed instructions on how to handle each kind of equations.
3. Many people have written to me since the publication of the first edition. They offered many useful and constructive suggestions. Their suggestions for extending some topics were honored in this text.
4. The fruitful evaluations, made by my students who used the first edition, provide useful input. My students' questions and concerns were addressed in this edition.

5. A new application chapter has been added to discuss a variety of scientific applications. Numerical and analytic treatments of linear and nonlinear integral equations are explained in this chapter to highlight the effectiveness of the traditional and the new methods.

6. The exact solutions of integral equations play a significant role in the proper understanding of the features and structures of the problem. For this reason, and based upon a vast request by readers, a solutions manual has been made that gives detailed explanations and illustrations for solving each problem of the second edition.

I would like to acknowledge the encouragement of my wife who supported me in all my endeavors, and in this edition and the accompanied solutions manual. I would also like to acknowledge our son and three daughters for their support and encouragement.

Saint Xavier University   Abdul-Majid Wazwaz
Chicago, IL 60655   e-mail: wazwaz@sxu.edu
Summer 2015

Preface to the First Edition

Engineering, physics and mathematics students, both advanced undergraduate and beginning graduate, need an integral equations textbook that simply and easily introduces the material. They also need a textbook that embarks upon their already acquired knowledge of regular integral calculus and ordinary differential equations. Because of these needs, this textbook was created. From many years of teaching, I have found that the available treatments of the subject are abstract. Moreover, most of them are based on comprehensive theories such as topological methods of functional analysis, Lebesgue integrals and Green functions. Such methods of introduction are not easily accessible to those who have not yet had a background in advanced mathematical concepts. This book is especially designed for those who wish to understand integral equations without having the extensive mathematical background. In this fashion, this text leaves out abstract methods, comprehensive methods and advanced mathematical topics.

From my experience in teaching and in guiding related senior seminar projects for advanced undergraduate students, I have found that the material can indeed be taught in an accessible manner. Students have showed both a lot of motivation and capability to grasp the subject once the abstract theories and difficult theorems were omitted. In my approach to teaching integral equations, I focus on easily applicable techniques and I do not emphasize such abstract methods as existence, uniqueness, convergence and Green functions. I have translated my means of introducing and fully teaching this subject into this text so that the intended user can take full advantage of the easily presented and explained material.

I have also introduced and made full use of some recent developments in this field.

The book consists of six chapters, each being divided into sections. In each chapter the equations are numbered consecutively and distinctly from

xiii

other chapters. Several examples are introduced in each section, and a large number of exercises are included to give the students a constructive insight through the material and to provide them with useful practice.

In this text, we were mainly concerned with linear integral equations, mostly of the second kind. The first chapter introduces classifictions of integral equations and necessary techniques to convert differential equations to integral equations or vice versa. The second chapter deals with linear Fredholm integral equations and the reliable techniques, supported by the new developments, to handle this style of equations. The linear Volterra integral equations are handled, using the recent developed techniques besides standard ones, in Chapter 3. The topic of integro-differential equations has been handled in Chapter 4 and reliable techniques were implemented to handle the essential link between differential and integral operators. The fifth chapter introduces the treatment of the singular and the weakly singular Volterra type integral equations. The sixth chapter deals with the nonlinear integral equations. This topic is difficult to study. However, recent schemes have been developed which show improvements over existing techniques and allow this topic to be far more easily accessible for specific cases. A large number of nonlinear integral examples and exercises are investigated.

Throughout the text, examples are provided to clearly and throughly introduce the new material in a clear and absorbable fashion. Many exercises are provided to give the new learner a chance to build his confidence and ease with the newly learned material. The exercises increase in complexity, to challenge the student.

Finally, the text has three useful Appendices. These Appendices provide the user with the integral forms, Maclaurin series and other related materials which are needed to be used in the exercises.

The author would highly appreciate any note concerning any error found.

Chicago, IL 1995   Abdul-Majid Wazwaz

Chapter 1

Introductory Concepts

1.1 Definitions

An integral equation is an equation in which the unknown function $u(x)$ to be determined appears under the integral sign. A typical form of an integral equation in $u(x)$ is of the form

$$u(x) = f(x) + \int_{\alpha(x)}^{\beta(x)} K(x, t)\, u(t)\, dt, \tag{1}$$

where $K(x, t)$ is called the kernel of the integral equation, and $\alpha(x)$ and $\beta(x)$ are the limits of integration. In (1), it is easily observed that the unknown function $u(x)$ appears under the integral sign as stated above, and out of the integral sign in most other cases as will be discussed later. It is important to point out that the kernel $K(x, t)$ and the function $f(x)$ in (1) are given in advance. Our goal is to determine $u(x)$ that will satisfy (1), and this may be achieved by using different techniques that will be discussed in the forthcoming chapters. The primary concern of this text will be focused on introducing these methods and techniques supported by illustrative and practical examples.

Integral equations arise naturally in physics, chemistry, biology and engineering applications modelled by initial value problems for a finite interval $[a, b]$. More details about the sources and origins of integral equations can be found in [12] and [14]. In the following example we will discuss how an initial value problem will be converted to the form of an integral equation.

Example 1. Consider the initial value problem

$$u'(x) = 2xu(x), \quad x \geq 0, \tag{2}$$

1

subject to the initial condition

$$u(0) = 1. \tag{3}$$

The equation (2) can be easily solved by using the method of separation of variables; where by using the initial condition (3), the solution

$$u(x) = e^{x^2}, \tag{4}$$

is easily obtained. However, integrating both sides of (2) with respect to x from 0 to x and using the initial condition, Eq. (3) yields the following

$$\int_0^x u'(t)\, dt = \int_0^x 2tu(t)\, dt, \tag{5}$$

or equivalently

$$u(x) = 1 + \int_0^x 2tu(t)\, dt, \tag{6}$$

obtained by integrating the left hand side of (5) and by using the given initial condition (3). Comparing (6) with (1) we find that $f(x) = 1$ and the kernel $K(x,t) = 2t$.

We will further discuss the algorithms of converting initial value problems and boundary value problems in detail to equivalent integral equations in the forthcoming sections. As stated above, our task is to determine the unknown function $u(x)$ that appears under the integral sign as in (1) and (6) and that will satisfy the given integral equation.

We further point out that integral equations as (1) and (6) are called *linear* integral equations. This classification is used if the unknown function $u(x)$ under the integral sign occurs linearly i.e. to the first power. However, if $u(x)$ under the integral sign is replaced by a nonlinear function in $u(x)$, such as $u^2(x)$, $\cos u(x)$, $\cosh u(x)$ and $e^{u(x)}$, *etc.*, the integral equation is called in this case a *nonlinear* integral equation.

1.2   Classification of Linear Integral Equations

The most frequently used linear integral equations fall under two main classes namely Fredholm and Volterra integral equations. However, in this text we will distinguish four more related types of linear integral equations in addition to the two main classes. In what follows, we will give a list of the Fredholm and Volterra integral equations, and the four related types:
1. Fredholm integral equations

2. Volterra integral equations
3. Integro-differential equations
4. Singular integral equations
5. Volterra-Fredholm integral equations
6. Volterra-Fredholm integro-differential equations

In the following we will outline the basic definitions and properties of each type.

1.2.1 Fredholm Linear Integral Equations

The standard form of Fredholm linear integral equations, where the limits of integration a and b are constants, are given by the form

$$\phi(x)\,u(x) = f(x) + \lambda \int_a^b K(x,t)u(t)\,dt,\ a \le x,\,t \le b, \qquad (7)$$

where the kernel of the integral equation $K(x,t)$ and the function $f(x)$ are given in advance, and λ is a parameter. The equation (7) is called *linear* because the unknown function $u(x)$ under the integral sign occurs linearly, i.e. the power of $u(x)$ is one.

The value of $\phi(x)$ will give the following kinds of Fredholm linear integral equations:

1. When $\phi(x) = 0$, Eq. (7) becomes

$$f(x) + \lambda \int_a^b K(x,t)u(t)\,dt = 0, \qquad (8)$$

and the integral equation is called Fredholm integral equation of the first kind.

2. When $\phi(x) = 1$, Eq. (7) becomes

$$u(x) = f(x) + \lambda \int_a^b K(x,t)u(t)dt, \qquad (9)$$

and the integral equation is called Fredholm integral equation of the second kind. In fact, the equation (9) can be obtained from (7) by dividing both sides of (7) by $\phi(x)$ provided that $\phi(x) \ne 0$.

In summary, the Fredholm integral equation is of the first kind if the unknown function $u(x)$ appears only under the integral sign. However, the Fredholm integral equation is of the second kind if the unknown function $u(x)$ appears inside and outside the integral sign.

1.2.2 Volterra Linear Integral Equations

The standard form of Volterra linear integral equations, where the limits of integration are functions of x rather than constants, are of the form

$$\phi(x)\,u(x) = f(x) + \lambda \int_a^x K(x,t)u(t)\,dt, \tag{10}$$

where the unknown function $u(x)$ under the integral sign occurs linearly as stated before. It is worth noting that (10) can be viewed as a special case of Fredholm integral equation when the kernel $K(x,t)$ vanishes for $t > x$, x is in the range of integration $[a,b]$.

As in Fredholm equations, Volterra integral equations fall under two kinds, depending on the value of $\phi(x)$, namely:

1. When $\phi(x) = 0$, Eq. (10) becomes

$$f(x) + \lambda \int_a^x K(x,t)u(t)\,dt = 0, \tag{11}$$

   and in this case the integral equation is called Volterra integral equation of the first kind.

2. When $\phi(x) = 1$, Eq. (10) becomes

$$u(x) = f(x) + \lambda \int_a^x K(x,t)u(t)dt, \tag{12}$$

   and in this case the integral equation is called Volterra integral equation of the second kind.

Examining the equations (7)–(12) carefully, the following remarks can be concluded:

In summary, the Volterra integral equation is of the first kind if the unknown function $u(x)$ appears only under the integral sign. However, the Volterra integral equation is of the second kind if the unknown function $u(x)$ appears inside and outside the integral sign.

Remarks

Examining the equations (7)–(12) carefully, the following remarks can be concluded.

1. **The structure of Fredholm and Volterra equations:**

   The unknown function $u(x)$ appears linearly only under the integral sign in linear Fredholm and Volterra integral equations of the first kind. However, the unknown function $u(x)$ appears linearly inside the integral sign and outside the integral sign as well in the second kind of both linear Fredholm and Volterra integral equations.

2. **The limits of integration:**

   In Fredholm integral equations, the integral is taken over a finite interval with fixed limits of integration. However, in Volterra integral equations, at least one limit of the range of integration is a variable, and the upper limit is the most commonly used with a variable limit.

3. **The origins of integral equations:**

   It is important to note that integral equations arise in engineering, physics, chemistry, and biology problems [12] and [14]. Further, integral equations arise as representation forms of differential equations. Furthermore, Fredholm and Volterra integral equations arise from different origins and applications, such as boundary value problems as in Fredholm equations, and from initial value problems as in Volterra equations. Based on the fact that integral equations arise from distinct origins, different techniques and approaches will be used to determine the solution of each type of integral equations.

4. **The linearity property:**

   As indicated before, the unknown function $u(x)$ in Fredholm and Volterra integral equations (9) and (12) occurs to the first power wherever it exists. However, nonlinear Fredholm and Volterra integral equations arise if $u(x)$ is replaced by a nonlinear function $F(u(x))$, such as $u^2(x), u^3(x), e^{u(x)}$ and so on. The following are examples of nonlinear integral equations:

   $$u(x) = f(x) + \lambda \int_a^x K(x,t)u^2(t)dt, \tag{13}$$

   $$u(x) = f(x) + \lambda \int_a^x K(x,t)e^{u(t)}dt, \tag{14}$$

   $$u(x) = f(x) + \lambda \int_0^1 K(x,t)\sin(u(t))dt, \tag{15}$$

   where the linear function $u(x)$ in (1) has been replaced by the nonlinear functions $u^2(t)$, $e^{u(t)}$ and $\sin(u(t))$ respectively.

5. **The homogeneity property:**

   If we set $f(x) = 0$ in Fredholm or Volterra integral equation of the second kind given by (9) and (12), the resulting equation is called a *homogeneous* integral equation, otherwise it is called *nonhomogeneous* or *inhomogeneous* integral equation.

6. **The singular behavior of the integral equation:**

   An integral equation is called singular if the integration is improper. This usually occurs if the interval of integration is *infinite*, or if the kernel becomes *unbounded* at one or more points of the interval of consideration a $\leq t \leq$ b. Singular integral equations will be defined later. However, the methods to handle singular integral equations will be introduced in Chapter 6.

It is worth noting that four other types of integral equations, related to the two main classes Fredholm and Volterra integral equations arise in many science and engineering applications. In the following, we introduce these significant equations as distinct types.

1.2.3   Integro-Differential Equations

Volterra, in the early 1900, studied the population growth, where new type of equations have been developed and was termed as integro-differential equations. In this type of equations, the unknown function $u(x)$ occurs in one side as an *ordinary derivative*, and appears on the other side under the *integral sign*. Several phenomena in physics and biology [14] and [20] give rise to this type of integro-differential equations. Further, we point out that an integro-differential equation can be easily observed as an intermediate stage when we convert a differential equation to an integral equation as will be discussed later in the coming sections.

The following are examples of integro-differential equations:

$$u^{''}(x) \;=\; -x + \int_0^x (x-t)u(t)dt, \quad u(0)=0,\; u'(0)=1, \qquad (16)$$

$$u^{'}(x) \;=\; -\sin x - 1 - + \int_0^x u(t)dt, \quad u(0)=1, \qquad (17)$$

$$u^{'}(x) \;=\; 1 - \frac{1}{3}x + \int_0^1 xtu(t)dt, \quad u(0)=1. \qquad (18)$$

Equations (16)–(17) are Volterra *integro-differential* equations, and (18) is a Fredholm *integro-differential* equation. This classification has been concluded as a result to the limits of integration. The solution for integro-differential equations will be established using in particular the most recent developed techniques. The integro-differential equations will be discussed extensively in Chapters 4 and 5.

1.2.4 Singular Integral Equations

The integral equation of the first kind

$$f(x) = \lambda \int_{\alpha(x)}^{\beta(x)} K(x,t)\, u(t) dt, \qquad (19)$$

or the integral equation of the second kind

$$u(x) = f(x) + \lambda \int_{\alpha(x)}^{\beta(x)} K(x,t)\, u(t) dt, \qquad (20)$$

is called *singular* if the lower limit, the upper limit or both limits of integration are *infinite*. In addition, the equation (19) or (20) is also called a singular integral equation if the kernel $K(x,t)$ becomes *infinite* at one or more points in the domain of integration. Examples of the first type of *singular* integral equations are given by the following examples:

$$u(x) = 2x + 6 \int_0^\infty \sin(x-t) u(t) dt, \qquad (21)$$

$$u(x) = x + \frac{1}{3} \int_{-\infty}^0 \cos(x+t) u(t) dt, \qquad (22)$$

$$u(x) = 1 + x^2 + \frac{1}{6} \int_{-\infty}^\infty (x+t) u(t) dt, \qquad (23)$$

where the singular behavior in these examples has resulted from the range of integration becoming *infinite*.

Examples of the second kind of *singular* integral equations are given by

$$x^2 = \int_0^x \frac{1}{\sqrt{x-t}} u(t) dt, \qquad (24)$$

$$x = \int_0^x \frac{1}{(x-t)^\alpha} u(t) dt, \quad 0 < \alpha < 1, \qquad (25)$$

$$u(x) = 1 - 2\sqrt{x} - \int_0^x \frac{1}{\sqrt{x-t}} u(t) dt, \qquad (26)$$

where the singular behavior in this kind of equations has resulted from the kernel $K(x,t)$ becoming *infinite* as $t \to x$.

It is important to note that integral equations similar to examples (24) and (25) are called Abel's integral equation and generalized Abel's integral equation respectively. Moreover these types of singular integral equations are among the earliest integral equations established by the Norwegian mathematician Niels Abel in 1823. Singular equations similar to example (26) are called the weakly-singular second-kind Volterra type integral equations. This type of equations usually arises in science and engineering applications like heat conduction, super-fluidity and crystal growth. The singular integral equations and the methods to handle it will be discussed in Chapter 6.

1.2.5   Volterra-Fredholm Integral Equations

The Volterra-Fredholm integral equation, which is a combination of disjoint Volterra and Fredholm integrals, appears in one integral equation. The Volterra-Fredholm integral equations arise from the modelling of the spatio-temporal development of an epidemic, from boundary value problems and from many physical and chemical applications. The standard form of the Volterra-Fredholm integral equation reads

$$u(x) = f(x) + \int_0^x K_1(x,t)u(t)dt + \int_a^b K_2(x,t)u(t)\,dt \qquad (27)$$

where $K_1(x,t)$ and $K_2(x,t)$ are the kernels of the equation.

Examples of the Volterra-Fredholm integral equations are

$$u(x) = 2x - \int_0^x (x-t)u(t)dt + \int_0^{\frac{\pi}{2}} xu(t)\,dt \qquad (28)$$

and

$$u(x) = \sin x - \cos x - \int_0^x u(t)dt + \int_0^{\frac{\pi}{2}} u(t)\,dt. \qquad (29)$$

Notice that the unknown function $u(x)$ appears inside the Volterra and Fredholm integrals and outside both integrals.

1.2.6   Volterra-Fredholm Integro-Differential Equations

The Volterra-Fredholm integro-differential equation, which is a combination of disjoint Volterra and Fredholm integrals and differential operator, may appear in one integral equation. The Volterra-Fredholm integro-differential equations arise from many physical and chemical applications similar to the Volterra-Fredholm integral equations. The standard form of the Volterra-Fredholm integro-differential equation reads

$$u^{(n)}(x) = f(x) + \int_0^x K_1(x,t)u(t)dt + \int_a^b K_2(x,t)u(t)\,dt \qquad (30)$$

where $K_1(x,t)$ and $K_2(x,t)$ are the kernels of the equation, and is the order of the ordinary derivative of $u(x)$. Notice that because this kind of equations contain ordinary derivatives, then initial conditions should be prescribed depending on the order of the derivative involved.

Examples of the Volterra-Fredholm integro-differential equations are

$$u'(x) = 1 + \int_0^x (x - t)u(t)dt + \int_0^1 xtu(t)\,dt, u(0) = 1 \tag{31}$$

and

$$u''(x) = -x - \frac{1}{6}x^3 + \int_0^x u(t)dt + \int_{-\pi}^\pi xu(t)\,dt, u(0) = 0, u'(0) = 2. \tag{32}$$

Notice that the unknown function $u(x)$ appears inside the Volterra and Fredholm integrals and outside both integrals.

In closing this section, we illustrate the classifications and the basic concepts that were discussed earlier by the following examples.

Example 1. Classify the following integral equation

$$u(x) = x - \frac{1}{6}x^3 + \int_0^x (x - t)u(t)dt, \tag{33}$$

as Fredholm or Volterra integral equation, *linear* or *nonlinear* and *homogeneous* or *nonhomogeneous*.

Note that the upper limit of the integral is x and the function $u(x)$ appears twice. This indicates that the equation (33) is a Volterra integral equation of the second kind. The equation (33) is *linear* since the unknown function $u(t)$ appears linearly inside and outside the integral sign. The presence of the function $f(x) = x - \frac{1}{3}x^3$ classifies the equation as a *nonhomogeneous* equation.

Example 2. Classify the following integral equation

$$u(x) = \frac{1}{2} + x - \int_0^1 (x - t)u^2(t)dt, \tag{34}$$

as Fredholm or Volterra integral equation, *linear* or *nonlinear* and *homogeneous* or *nonhomogeneous*.

The limits of integration are constants and the function $u(x)$ appears twice, therefore the equation (34) is a Fredholm integral equation of the second kind. Further, the unknown function appears under the integral sign with power two indicating the equation is a *nonlinear* equation. The nonhomogeneous part $f(x)$ appears in the equation showing that it is a *nonhomogeneous* equation.

Example 3. Classify the following equation

$$u'(x) = 1 - \frac{1}{3}x^3 + \int_0^x tu(t)dt, \quad u(0) = 0, \tag{35}$$

as Fredholm or Volterra integro-differential equation, and *linear* or *nonlinear*.

It is easily seen that (35) includes differential and integral operators, and by noting that the upper limit of the integral is a variable, we conclude that (35) is a Volterra integro-differential equation. Moreover, the equation is *linear* since $u(x)$ and $u'(x)$ appear linearly in the equation.

We finally discuss the following example.

Example 4. Discuss the type, linearity and homogeneity of the following equation

$$u(x) = 1 + \int_0^x tu(t)dt + \int_0^\pi \sin(u(t))u(t)\, dt. \tag{36}$$

This equation combines Volterra and Fredholm integrals, hence it is Volterra–Fredholm integral equation. It is nonlinear because of the term $\sin(u(t))$. It is nonhomogeneous because $f(x) = 1$.

We point out that linear Fredholm integral equations, linear Volterra integral equations, integro-differential equations, singular integral equations, Volterra-Fredholm integral and integro-differential equations will be discussed in the forthcoming chapters. The Volterra and Fredholm nonlinear integral equations will also be briefly discussed. The recent developed methods, that proved its effectiveness and reliability, will be applied to all types of integral equations. In other words, we will use the Adomian decomposition method (ADM) and the variational method (VIM) for handling these chapters. Some of the traditional methods will also be used so that newly developed methods and traditional methods complement each other.

Exercises 1.2

In exercises 1-10, classify each of the following integral equations as Fredholm or Volterra integral equation, *linear* or *nonlinear*, and *homogeneous* or *nonhomogeneous*:

(1) $u(x) = x + \int_0^1 xtu(t)\, dt$

(2) $u(x) = 1 + x^2 + \int_0^x (x - t)u(t)\, dt$

(3) $u(x) = e^x + \int_0^x tu^2(t)\, dt$

(4) $u(x) = \int_0^1 (x - t)^2 u(t)\, dt$

(5) $u(x) = \dfrac{2}{3}x + \int_0^1 xtu(t)\, dt$

(6) $u(x) = \dfrac{3}{4}x + \dfrac{1}{5} + \int_0^1 (x - t)^3 u(t)\, dt$

(7) $u(x) = 1 + \dfrac{x}{4} \int_0^1 \dfrac{1}{x+t} \dfrac{1}{u(t)}\, dt$

(8) $u(x) = \dfrac{1}{2}\cos x + \dfrac{1}{2} \int_0^{\frac{\pi}{2}} \cos x\, u(t)\, dt$

(9) $u(x) = 1 + \int_0^x (x - t)^2 u^2(t)\, dt$

(10) $u(x) = 1 - \int_0^x (x - t)\, u(t)\, dt$

In exercises 11–15, classify the following integro-differential equations as Fredholm integro-differential equation or Volterra integro-differential equation. Also determine whether the equation is *linear* or *nonlinear*.

(11) $u'(x) = 1 + \int_0^x e^{-2t} u^3(t)\, dt, \quad u(0) = 1$

(12) $u'(x) = 1 - \dfrac{1}{3}x + \int_0^1 xtu(t)\, dt, \quad u(0) = 0$

(13) $u''(x) = \dfrac{1}{2}x^2 - \int_0^x (x - t) u^3(t)\, dt, \quad u(0) = 1, u'(0) = 0$

(14) $u'''(x) = \sin x - x + \int_0^{\frac{\pi}{2}} xtu'(t)\, dt,$

$u(0) = 1, u'(0) = 0, u''(0) = -1$

(15) $u''''(x) = -\dfrac{1}{12}x^4 + \int_0^x (x - t) u(t)\, dt,$

$u(0) = u'(0) = 0, u''(0) = 2$

In exercises 16–20, integrate both sides of each of the following differential equations once from 0 to x, and use the given initial condition to convert to a corresponding integral equation or integro-differential equation. (Follow Example 1.)

(16) $u'(x) = 4u(x), \quad u(0) = 1.$
(17) $u'(x) = 3x^2 u(x), \quad u(0) = 1.$
(18) $u'(x) = u^2(x), \quad u(0) = 4.$
(19) $u''(x) = 4xu^2(x), \quad u(0) = 2, u'(0) = 1.$
(20) $u''(x) = 2xu(x), \quad u(0) = 0, u'(0) = 1.$

In exercises 21–22, discuss the type, linearity and homogeneity of the following equations

(21) $u(x) = x + \int_0^x \tan(u(t))u(t)dt - \int_0^1 (x - t)u(t)\,dt$

(22) $u'(x) = 1 + \int_0^x (x - t)u(t)dt + \int_{-1}^1 (xtu(t)\,dt, u(0) = 1$

(23) $u'(x) = \sin x + \int_0^x tu(t)dt + \int_0^1 t^2 u^2(t)\,dt, u(0) = 0$

(24) $u(x) = 1 + x + \int_0^x \frac{u^2(t)}{\sqrt{x-t}}\,dt$

1.3   Solution of an Integral Equation

A solution of an integral equation or an integro-differential equation on the interval of integration is a function $u(x)$ such that it satisfies the given equation. In other words, if the given solution is substituted in the right hand side of the equation, the output of this direct substitution must yield the left hand side, i.e. we should verify that the given function $u(x)$ satisfies the integral equation or the integro-differential equation under discussion. This important concept will be illustrated first by examining the following examples.

Example 1. Show that $u(x) = e^x$ is a solution of the Volterra integral equation

$$u(x) = 1 + \int_0^x u(t)dt. \tag{37}$$

Substituting $u(x) = e^x$ in the right hand side (RHS) of (37) yields

$$
\begin{aligned}
RHS &= 1 + \int_0^x e^t dt \\
&= 1 + [e^t]_0^x \\
&= e^x = u(x) \\
&= LHS.
\end{aligned}
\tag{38}
$$

Example 2. Show that $u(x) = x$ is a solution of the following Fredholm integral equation

$$u(x) = \frac{5}{6}x - \frac{1}{9} + \frac{1}{3}\int_0^1 (x + t)u(t)dt. \tag{39}$$

Substituting $u(x) = x$ in the right hand side of (39) we obtain

$$
\begin{aligned}
RHS &= \frac{5}{6}x - \frac{1}{9} + \frac{1}{3}\int_0^1 (x+t)u(t)dt \\
&= \frac{5}{6}x - \frac{1}{9} + \frac{1}{3}[\frac{xt^2}{2} + \frac{t^3}{3}]_0^1 \\
&= x = u(x) \\
&= LHS.
\end{aligned}
\tag{40}
$$

Example 3. Show that $u(x) = x$ is a solution of the following Fredholm integro-differential equation

$$
u'(x) = \frac{2}{3} + \int_0^1 tu(t)dt.
\tag{41}
$$

Substituting $u(x) = x$ in the right hand side of (41) we obtain

$$
\begin{aligned}
RHS &= \frac{2}{3} + \int_0^1 t^2\, dt \\
&= \frac{2}{3} + [\frac{t^3}{3}]_0^1 \\
&= 1 = u'(x) \\
&= LHS.
\end{aligned}
\tag{42}
$$

Example 4. Check if $u(x) = x + e^x$ is a solution of the following Fredholm integral equation

$$
u''(x) = e^x - \frac{4}{3}x + \int_0^1 xt\, u(t)\, dt, \, u(0) = 1, u'(0) = 2.
\tag{43}
$$

Substituting $u(x) = x + e^x$ in the right hand side of (43) we obtain

$$
\begin{aligned}
RHS &= e^x - \frac{4}{3}x + + \int_0^1 xt(t + e^t)\, dt \\
&= e^x = u''(x) \\
&= LHS.
\end{aligned}
\tag{44}
$$

Example 5. If $u(x) = e^{-x^2}$ is a solution of the following Volterra integral equation

$$
u(x) = 1 - \alpha \int_0^x t\, u(t)dt.
\tag{45}
$$

Find α.

Substituting $u(x) = e^{-x^2}$ into both sides of (45) we obtain

$$e^{-x^2} = 1 - \frac{\alpha}{2} + \frac{\alpha}{2}e^{-x^2}. \tag{46}$$

Solving for α gives

$$\alpha = 2. \tag{47}$$

Example 6. $u(x) = x^2 + x^3$ is a solution of the Fredholm integral equation

$$u(x) = x^3 - x^2 - 2x + \alpha \int_{-1}^{1} (xt^2 + x^2t)u(t)dt. \tag{48}$$

Find α.

Substituting $u(x) = x^2 + x^3$ into both sides of (48) we obtain

$$x^2 + x^3 = x^3 - x^2 - 2x + \alpha(\frac{2}{5}x^2 + \frac{2}{5}x). \tag{49}$$

Solving for α gives

$$\alpha = 5. \tag{50}$$

Three useful remarks can be made with respect to the concept of the solution of an integral equation or an integro-differential equation. First, the question of *existence* of a solution and the question of *uniqueness* of a solution, that usually we discuss in differential equations and integral equations will be left for further studies.

We next remark that if a solution exists for an integral equation or an integro-differential equation, it is important to note that this solution may be given in a closed form expressed in terms of elementary functions, such as a polynomial, exponential, trigonometric or hyperbolic function, similar to the solutions given in the previous examples. However, it is not always possible to obtain the solution in a closed form, but instead the obtained solution may be expressible in a series form. The solution obtained in a series form is usually used for numerical approximations, and in this case the more terms we obtain the better accuracy level we achieve.

It is important to illustrate the difference between the two expressible forms, the exact solution in a closed form and the approximant solution in a series form. Considering Example 1 above, we note that the exact solution is given in a closed form by the exponential function $u(x) = e^x$. However, it will be shown later that the solution of the integral equation

$$u(x) = 1 + \frac{1}{4} \int_{0}^{x} xu(t)dt, \tag{51}$$

is given by the series form

$$u(x) = 1 + \frac{1}{4}x^2 + \frac{1}{48}x^4 + \frac{1}{960}x^6 + \cdots, \tag{52}$$

where we can easily observe that it is difficult to express the series (52) in an equivalent closed form. As indicated earlier, the series obtained can be used for numerical purposes, and to achieve the highly desirable accuracy we should determine more terms in the series solution.

In closing our remarks, we consider the nonlinear Fredholm integral equation

$$u(x) = \frac{5}{6}x + \frac{1}{2}\int_0^1 xu^2(t)dt. \tag{53}$$

It was found that equation (53) has two real solutions given by

$$u(x) = x, \, 5x, \tag{54}$$

and this can be justified through direct substitution. The *uniqueness* concept is not applicable for this example and for many other nonlinear problems. The nonlinear problems will be examined briefly in forthcoming chapters. The uniqueness criteria for nonlinear problems is justified for only specific problems under specific conditions. Generally speaking, nonlinear problems, differential and integral, give more than one solutions.

It is useful to point out that our main concern in this text will be on the linear integral equations and the linear integro-differential equations only. The nonlinear Fredholm and the nonlinear Volterra integral equations will be examined as well. In addition, we will focus our study on equations with closed form solutions as in Examples 1 and 2. Other cases that may lead to a series solution will be investigated as well, supported by the development of the reliable techniques that will be discussed later. Moreover, nonlinear integral equations will be investigated in its simplest forms. The recent developed methods presented powerful techniques, and therefore these methods have been carried out with promising results in linear and nonlinear equations.

Exercises 1.3

In exercises 1–10, verify that the given function is a solution of the corresponding integral or integro-differential equation:

(1) $u(x) = x + \int_0^{\frac{1}{4}} u(t)dt, \quad u(x) = x + \frac{1}{24}$

(2) $u(x) = \frac{2}{3}x + \int_0^1 xtu(t)dt, \quad u(x) = x$

(3) $u(x) = x + \displaystyle\int_0^1 xtu^2(t)dt, \quad u(x) = 2x$

(4) $u(x) = x - \displaystyle\int_0^x (x - t)u(t)dt, \quad u(x) = \sin x$

(5) $u(x) = 2\cosh x - x\sinh x - 1 + \displaystyle\int_0^x tu(t)dt, \quad u(x) = \cosh x$

(6) $u(x) = x + \dfrac{1}{5}x^5 - \displaystyle\int_0^x tu^3(t)dt, \quad u(x) = x$

(7) $u'(x) = 2x - x^4 + \displaystyle\int_0^x 4tu(t)dt, \quad u(0) = 0, \quad u(x) = x^2$

(8) $u''(x) = x\cos x - 2\sin x + \displaystyle\int_0^x tu(t)dt,$

$\quad u(0) = 0, \ u'(0) = 1, \ u(x) = \sin x$

(9) $\displaystyle\int_0^x (x - t)^2 u(t)dt = x^3, \quad u(x) = 3$

(10) $\displaystyle\int_0^x (x - t)^{1/2} u(t)dt = x^{3/2}, \quad u(x) = \dfrac{3}{2}$

In exercises 11–14, find the unknown if the solution of each equation is given:

(11) $u(x) = 1 - \dfrac{1}{2}f(x) - \displaystyle\int_0^x (x - t)u(t)\,dt \quad u(x) = 2\cos x - 1$

(12) $u(x) = f(x) + \displaystyle\int_0^x (x - t)u(t)\,dt \quad u(x) = e^x$

(13) $u(x) = 1 - \alpha \displaystyle\int_0^x 3t^2 u(t)\,dt \quad u(x) = e^{-x^3}$

(14) $u(x) = f(x) - 1 + \displaystyle\int_0^{\frac{\pi}{2}} tu(t)\,dt \quad u(x) = \sin x$

1.4   Converting Volterra Equation to an ODE

In this section we will present the technique that converts Volterra integral equations of the second kind to equivalent differential equations. This may be easily achieved by applying the important Leibniz Rule for differentiating an integral. It seems reasonable to review the basic outline of the rule.

1.4.1 Differentiating Any Integral: Leibniz Rule

To differentiate the integral $\int_{\alpha(x)}^{\beta(x)} G(x,t)dt$ with respect to x, we usually apply the useful Leibniz rule given by:

$$\frac{d}{dx} \int_{\alpha(x)}^{\beta(x)} G(x,t)dt = G(x,\beta(x)) \frac{d\beta}{dx} - G(x,\alpha(x)) \frac{d\alpha}{dx} + \int_{\alpha(x)}^{\beta(x)} \frac{\partial G}{\partial x} dt,$$

(55)

where $G(x,t)$ and $\frac{\partial G}{\partial x}$ are continuous functions in the domain D in the xt-plane that contains the rectangular region R, $a \leq x \leq b$, $t_0 \leq t \leq t_1$, and the limits of integration $\alpha(x)$ and $\beta(x)$ are defined functions having continuous derivatives for $a < x < b$. We note that Leibniz rule is usually presented in most calculus books, and our concern will be on using the rule rather than its theoretical proof. The following examples are illustrative and will be mostly used in the coming approach that will be used to convert Volterra integral equations to differential equations.

Example 1. Find $\dfrac{d}{dx} \displaystyle\int_0^x (x-t)^2 u(t)\, dt$.

In this example, $\alpha(x) = 0$, $\beta(x) = x$, hence $\alpha'(x) = 0$, $\beta'(x) = 1$ and $\dfrac{\partial G}{\partial x} = 2(x-t)u(t)$. Using Leibniz rule (55), we find

$$\frac{d}{dx} \int_0^x (x-t)^2 u(t)\, dt = \int_0^x 2(x-t)u(t)\, dt.$$

(56)

Example 2. Find $\dfrac{d}{dx} \displaystyle\int_0^x (x-t)u(t)\, dt$.

In this example, $\alpha(x) = 0$, $\beta(x) = x$, hence $\alpha'(x) = 0$, $\beta'(x) = 1$, and $\frac{\partial G}{\partial x} = u(t)$. Using Leibniz rule (55), we find

$$\frac{d}{dx} \int_0^x (x-t)u(t)\, dt = \int_0^x u(t)dt.$$

(57)

Example 3. Find $\dfrac{d}{dx} \displaystyle\int_0^x u(t)\, dt$.

Proceeding as before, we find that

$$\frac{d}{dx} \int_0^x u(t)\, dt = u(x).$$

(58)

We now turn to our main goal to convert a Volterra integral equation to an equivalent differential equation. This can be easily achieved by differentiating both sides of the integral equation, noting that Leibniz rule should be used in differentiating the integral as stated above. The differentiating process should be continued as many times as needed until we obtain a pure differential equation with the integral sign removed. Moreover, the initial conditions needed can be obtained by substituting $x = 0$ in the integral equation and the resulting integro-differential equations as will be shown.

We are now ready to give the following illustrative examples.

Example 4. Find the initial value problem equivalent to the Volterra integral equation

$$u(x) = 1 + \int_0^x u(t)dt. \tag{59}$$

Differentiating both sides of the integral equation and using Leibniz rule we find

$$u'(x) = u(x). \tag{60}$$

The initial condition can be obtained by substituting $x = 0$ into both sides of the integral equation; hence we find $u(0) = 1$. Consequently, the corresponding initial value problem of first order is given by

$$u'(x) - u(x) = 0, \quad u(0) = 1. \tag{61}$$

Example 5. Convert the following Volterra integral equation to an initial value problem

$$u(x) = x + \int_0^x (t - x)u(t)dt. \tag{62}$$

Differentiating both sides of the integral equation we obtain

$$u'(x) = 1 - \int_0^x u(t)dt. \tag{63}$$

We differentiate both sides of the resulting integro-differential equation (63) to remove the integral sign, therefore we obtain

$$u''(x) = -u(x), \tag{64}$$

or equivalently

$$u''(x) + u(x) = 0. \tag{65}$$

The related initial conditions are obtained by substituting $x = 0$ in $u(x)$ and in $u'(x)$ in the equations above, and as a result we find $u(0) = 0$ and $u'(0) = 1$. Combining the above results yields the equivalent initial value problem of the second order given by

$$u''(x) + u(x) = 0, \ u(0) = 0, u'(0) = 1, \tag{66}$$

with constant coefficients that can be easily handled.

Example 6.   Find the initial value problem equivalent to the Volterra integral equation

$$u(x) = x^3 + \int_0^x (x - t)^2 u(t)\, dt. \tag{67}$$

Differentiating both sides of (67) three times we find

$$\begin{cases} u^{'}(x) &=& 3x^2 + 2\displaystyle\int_0^x (x - t)u(t)dt, \\[2mm] u^{''}(x) &=& 6x + 2\displaystyle\int_0^x u(t)dt, \\[2mm] u^{'''}(x) &=& 6 + 2u(x). \end{cases} \tag{68}$$

The proper initial conditions can be easily obtained by substituting $x = 0$ in $u(x)$, $u'(x)$ and $u''(x)$ in the obtained equations above. Consequently, we obtain the nonhomogeneous initial value problem of third order given by

$$u^{'''}(x) - 2u(x) = 6, \ \ u(0) = u^{'}(0) = u^{''}(0) = 0, \tag{69}$$

with constant coefficients that can be easily solved as an ordinary differential equation.

We point out here that the solution of initial value problems, that result from converting Volterra integral equations, will be discussed in Chapter 3.

Exercises 1.4

In exercise 1-4, find $\dfrac{d}{dx}$ for the given integrals by using Leibniz rule:

(1) $\displaystyle\int_0^x (x - t)^3 u(t)\, dt$

(2) $\displaystyle\int_x^{x^2} e^{xt}\, dt$

(3) $\displaystyle\int_0^x (x - t)^4 u(t)\, dt$

(4) $\displaystyle\int_x^{4x} \sin(x + t)\, dt$

In exercise 5-14, convert each of the Volterra integral equations to an equivalent initial value problem:

(5) $u(x) = 1 + x + \displaystyle\int_0^x (x - t)^2 u(t)\, dt$

(6) $u(x) = e^x - \int_0^x (x - t)u(t)\, dt$

(7) $u(x) = x + \int_0^x (x - t)u(t)\, dt$

(8) $u(x) = x - \cos x + \int_0^x (x - t)u(t)\, dt$

(9) $u(x) = 2 + 3x + 5x^2 + \int_0^x [1 + 2(x - t)]u(t)\, dt$

(10) $u(x) = -5 + 6x + \int_0^x (5 - 6x + 6t)u(t)\, dt$

(11) $u(x) = \tan x - \int_0^x u(t)\, dt,\ x < \pi/2$

(12) $u(x) = 1 + x + \dfrac{5}{2}x^2 + \int_0^x [3 + 6(x - t) - \dfrac{5}{2}(x - t)^2]u(t)\, dt$

(13) $u(x) = x^4 + x^2 + 2\int_0^x (x - t)^2 u(t)\, dt$

(14) $u(x) = x^2 + \dfrac{1}{6}\int_0^x (x - t)^3 u(t)\, dt$

1.5   Converting IVP to Volterra Equation

In this section, we will study the method that converts an initial value problem to an equivalent Volterra integral equation. Before outlining the method needed, we wish to recall the useful transformation formula

$$\int_0^x \int_0^{x_1} \int_0^{x_2} \cdots \int_0^{x_{n-1}} f(x_n)\, dx_n \cdots dx_1 = \frac{1}{(n-1)!}\int_0^x (x - t)^{n-1} f(t)\, dt,$$
(70)

that converts any multiple integral to a single integral. This is an essential and useful formula that will be employed in the method that will be used in the conversion technique. We point out that this formula appears in most calculus texts. For practical considerations, the formulas

$$\int_0^x \int_0^x f(t)\, dt\, dt = \int_0^x (x - t)\, f(t)\, dt,$$
(71)

and

$$\int_0^x \int_0^x \int_0^x f(t)\, dt\, dt\, dt = \frac{1}{2!}\int_0^x (x - t)^2\, f(t)\, dt,$$
(72)

are two special cases of the formula given above, and the mostly used formulas that will transform double and triple integrals respectively to a single integral for each. For simplicity reasons, we prove the first formula (71) that converts double integral to a single integral. Noting that the right hand side of (71) is a function of x allows us to set the equation

$$I(x) = \int_0^x (x - t) f(t) \, dt. \tag{73}$$

Differentiating both sides of (73), and using Leibniz rule we obtain

$$I'(x) = \int_0^x f(t) \, dt. \tag{74}$$

Integrating both sides of (74) from 0 to x, noting that $I(0) = 0$ from (73), we find

$$I(x) = \int_0^x \int_0^x f(t) \, dt dt. \tag{75}$$

Equating the right hand sides of (73) and (75) completes the proof for this special case. The proof for the conversion of the triple integral to a single integral given by (72) may be carried out in the same manner. The procedure of reducing multiple integral to a single integral will be illustrated by examining the following examples.

Example 1. Convert the following quadruple integral

$$I(x) = \int_0^x \int_0^x \int_0^x \int_0^x u(t) \, dt dt dt dt, \tag{76}$$

to a single integral.

Using the formula (70), noting that $n = 4$, we find

$$I(x) = \frac{1}{3!} \int_0^x (x - t)^3 \, u(t) \, dt, \tag{77}$$

the equivalent single integral.

Example 2. Convert the triple integral

$$I(x) = \int_0^x \int_0^x \int_0^x u(t) \, dt dt dt, \tag{78}$$

to a single integral.

Using the formula (70) yields

$$I(x) = \frac{1}{2!} \int_0^x (x - t)^2 u(t) \, dt, \tag{79}$$

the equivalent single integral.

Returning to the main goal of this section, we discuss the technique that will be used to convert an initial value problem to an equivalent Volterra integral equation. Without loss of generality, and for simplicity reasons, we apply this technique to a third order initial value problem given by

$$y'''(x) + p(x)y''(x) + q(x)y'(x) + r(x)y(x) = g(x) \tag{80}$$

subject to the initial conditions

$$y(0) = \alpha, y'(0) = \beta, y''(0) = \gamma, \quad \alpha, \ \beta \ \text{and} \ \gamma \ \text{are constants.} \tag{81}$$

The coefficient functions $p(x)$, $q(x)$ and $r(x)$ are analytic functions by assuming that these functions have Taylor expansions about the origin. Besides, we assume that $g(x)$ is continuous through the interval of discussion. To transform (80) into an equivalent Volterra integral equation, we first set

$$y'''(x) = u(x), \tag{82}$$

where $u(x)$ is a continuous function on the interval of discussion. Based on (82), it remains to find other relations for y and its derivatives as single integrals involving $u(x)$. This can be simply performed by integrating both sides of (82) from 0 to x where we find

$$y''(x) - y''(0) = \int_0^x u(t)dt, \tag{83}$$

or equivalently

$$y''(x) = \gamma + \int_0^x u(t)dt, \tag{84}$$

obtained upon using the initial condition $y''(0) = \gamma$. To obtain $y'(x)$ we integrate both sides of (84) from 0 to x to find that

$$y'(x) = \beta + \gamma x + \int_0^x \int_0^x u(t)dtdt. \tag{85}$$

Similarly we integrate both sides of (85) from 0 to x to obtain

$$y(x) = \alpha + \beta x + \frac{1}{2}\gamma x^2 + \int_0^x \int_0^x \int_0^x u(t)dtdtdt. \tag{86}$$

Using the conversion formulas (71) and (72), to reduce the double and triple integrals in (85) and (86) respectively to single integrals, we obtain

$$y'(x) = \beta + \gamma x + \int_0^x (x - t)u(t)dt, \tag{87}$$

and

$$y(x) = \alpha + \beta x + \frac{1}{2}\gamma x^2 + \frac{1}{2}\int_0^x (x-t)^2 u(t)dt, \tag{88}$$

respectively. Substituting (82), (84), (87) and (88) into (80) leads to the following Volterra integral equation of the second kind

$$u(x) = f(x) + \int_0^x K(x,t)u(t)dt, \tag{89}$$

where

$$K(x,t) = p(x) + q(x)(x-t) + \frac{1}{2!}r(x)(x-t)^2, \tag{90}$$

and

$$f(x) = g(x) - \left\{ \gamma p(x) + \beta q(x) + \alpha r(x) + \gamma x q(x) + r(x)\left(\beta x + \frac{1}{2}\gamma x^2\right) \right\}. \tag{91}$$

The following examples will be used to illustrate the above discussed technique.

Example 3. Convert the following initial value problem

$$y''' - 3y'' - 6y' + 5y = 0, \tag{92}$$

subject to the initial conditions

$$y(0) = y'(0) = y''(0) = 1, \tag{93}$$

to an equivalent Volterra integral equation.

As indicated before, we first set

$$y'''(x) = u(x). \tag{94}$$

Integrating both sides of (94) from 0 to x and using the initial condition $y''(0) = 1$ we find

$$y''(x) = 1 + \int_0^x u(t)dt. \tag{95}$$

Integrating (95) twice and using the proper initial conditions we find

$$y'(x) = 1 + x + \int_0^x \int_0^x u(t)dtdt \tag{96}$$

and

$$y(x) = 1 + x + \frac{1}{2}x^2 + \int_0^x \int_0^x \int_0^x u(t)dtdtdt. \tag{97}$$

Transforming the double and triple integrals in (96) and (97) to single integrals by using the formulas (71) and (72) we find

$$y'(x) = 1 + x + \int_0^x (x-t)u(t)dt, \tag{98}$$

and

$$y(x) = 1 + x + \frac{1}{2}x^2 + \frac{1}{2}\int_0^x (x-t)^2 u(t)dt. \tag{99}$$

Substituting (94), (95), (98) and (99) into (92) we find

$$u(x) = 4 + x - \frac{5}{2}x^2 + \int_0^x \left(3 + 6(x-t) - \frac{5}{2}(x-t)^2\right)u(t)dt, \tag{100}$$

the equivalent Volterra integral equation.

Example 4. Find the equivalent Volterra integral equation to the following initial value problem

$$y''(x) + y(x) = \cos x, \ y(0) = 0, \ y'(0) = 1. \tag{101}$$

Proceeding as before, we set

$$y''(x) = u(x). \tag{102}$$

Integrating both sides of (102) from 0 to x, using the initial condition $y'(0) = 1$ yields

$$y'(x) = 1 + \int_0^x u(t)dt. \tag{103}$$

Integrating (103), using the initial condition $y(0) = 0$ leads to

$$y(x) = x + \int_0^x \int_0^x u(t)\, dt dt, \tag{104}$$

or equivalently

$$y(x) = x + \int_0^x (x-t)u(t)\, dt, \tag{105}$$

upon using the conversion rule (71). Inserting (102) and (105) into (101) leads to the following required Volterra integral equation

$$u(x) = \cos x - x - \int_0^x (x-t)u(t)\, dt, \tag{106}$$

the equivalent Volterra integral equation.

As previously remarked, linear Volterra integral equations will be discussed extensively in Chapter 3. It is of interest to point out that the newly developed methods and the traditional methods will be introduced in that chapter.

Exercises 1.5

In exercises 1-3, convert each of the following first order initial value problem to a Volterra integral equation:

(1) $y' + y = 0$, $y(0) = 1$
(2) $y' - y = x$, $y(0) = 0$
(3) $y' + y = \sec^2 x$, $y(0) = 0$

In exercises 4-10, derive an equivalent Volterra integral equation to each of the following initial value problems of second order:

(4) $y'' + y = 0$, $y(0) = 1, y'(0) = 0$
(5) $y'' - y = 0$, $y(0) = 1, y'(0) = 1$
(6) $y'' + 5y' + 6y = 0$, $y(0) = 1$, $y'(0) = 1$
(7) $y'' + y' = 0$, $y(1) = 0$, $y'(1) = 1$
(8) $y'' + y' - 2y = 2x$, $y(0) = 0$, $y'(0) = 1$
(9) $y'' + y = \sin x$, $y(0) = 0$, $y'(0) = 0$
(10) $y'' - \sin x \, y' + e^x y = x$, $y(0) = 1$, $y'(0) = -1$

In exercises 11-15, convert each of the following initial value problems of higher order to an equivalent Volterra integral equation:

(11) $y''' - y'' - y' + y = 0$, $y(0) = 2$, $y'(0) = 0$, $y''(0) = 2$
(12) $y''' + 4y' = x$, $y(0) = 0$, $y'(0) = 0$, $y''(0) = 1$
(13) $y^{iv} + 2y'' + y = 3x + 4$, $y(0) = 0, y'(0) = 0, y''(0) = 1, y'''(0) = 1$
(14) $y^{iv} - y = 0$, $y(0) = 1$, $y'(0) = 0$, $y''(0) = -1, y'''(0) = 0$
(15) $y^{iv} + y'' = 2e^x$, $y(0) = 2$, $y'(0) = 2$, $y''(0) = 1, y'''(0) = 1$

1.6 Converting BVP to Fredholm Equation

So far we have discussed how an initial value problem can be transformed to an equivalent Volterra integral equation. In this section, we will present the technique that will be used to convert a boundary value problem to an equivalent Fredholm integral equation. The technique is similar to that discussed in the previous section with some exceptions that are related to the boundary conditions. It is important to point out here that the procedure of reducing boundary value problem to Fredholm integral equation is complicated and rarely used. The method is similar to the technique discussed above, that reduces initial value problem to Volterra integral equation, with the exception that we are given boundary conditions.

A special attention should be taken to define $y'(0)$, since it is not always given, as will be seen later. This can be easily determined from the resulting equations. It seems useful and practical to illustrate this method by applying it to an example rather than proving it.

Example 1. We want to derive an equivalent Fredholm integral equation to the following boundary value problem

$$y''(x) + y(x) = x, \ 0 < x < \pi, \tag{107}$$

subject to the boundary conditions

$$y(0) = 1, \ y(\pi) = \pi - 1. \tag{108}$$

We first set

$$y''(x) = u(x). \tag{109}$$

Integrating both sides of (109) from 0 to x gives

$$\int_0^x y''(t)dt = \int_0^x u(t)\, dt, \tag{110}$$

or equivalently

$$y'(x) = y'(0) + \int_0^x u(t)\, dt. \tag{111}$$

As indicated earlier, $y'(0)$ is not given in this boundary value problem. However, $y'(0)$ will be determined later by using the boundary condition at $x = \pi$.

Integrating both sides of (111) from 0 to x and using the given boundary condition at $x = 0$ we find

$$y(x) = 1 + xy'(0) + \int_0^x (x - t)u(t)\, dt, \tag{112}$$

upon converting the resulting double integral to a single integral as discussed before. It remains to evaluate $y'(0)$, and this can be obtained by substituting $x = \pi$ in both sides of (112) and using the boundary condition at $x = \pi$, hence we find

$$y(\pi) = 1 + \pi y'(0) + \int_0^\pi (\pi - t)u(t)\, dt. \tag{113}$$

Solving (113) for $y'(0)$ we obtain

$$y'(0) = \frac{1}{\pi}\left((\pi - 2) - \int_0^\pi (\pi - t)\, u(t)\, dt\right). \tag{114}$$

Substituting (114) for $y'(0)$ into (112) yields

$$y(x) = 1 + \frac{x}{\pi}\left((\pi - 2) - \int_0^\pi (\pi - t)u(t)dt\right) + \int_0^x (x - t)u(t)\, dt. \tag{115}$$

Substituting (109) and (115) into (107) we get

$$\begin{aligned} u(x) &= x - 1 - \frac{x}{\pi}\left((\pi - 2) - \int_0^\pi (\pi - t)u(t)dt\right) \\ &\quad - \int_0^x (x - t)u(t)\, dt. \end{aligned} \tag{116}$$

The following identity

$$\int_0^\pi (.) = \int_0^x (.) + \int_x^\pi (.), \tag{117}$$

will carry the equation (116) to

$$\begin{aligned} u(x) &= x - 1 - \frac{x}{\pi}(\pi - 2) + \frac{x}{\pi}\int_0^x (\pi - t)u(t)dt \\ &\quad + \frac{x}{\pi}\int_x^\pi (\pi - t)u(t)\, dt - \int_0^x (x - t)u(t)dt, \end{aligned} \tag{118}$$

or equivalently, after performing simple calculations and adding integrals with similar limits

$$u(x) = \frac{2x - \pi}{\pi} - \int_0^x \frac{t(x - \pi)}{\pi}u(t)dt - \int_x^\pi \frac{x(t - \pi)}{\pi}u(t)dt. \tag{119}$$

Consequently, the desired Fredholm integral equation of the second kind is given by

$$u(x) = \frac{2x - \pi}{\pi} - \int_0^\pi K(x,t)u(t)\, dt, \tag{120}$$

where the kernel $K(x,t)$ is defined by

$$K(x,t) = \begin{cases} \dfrac{t(x - \pi)}{\pi} & \text{for } 0 \le t \le x \\[2mm] \dfrac{x(t - \pi)}{\pi} & \text{for } x \le t \le \pi. \end{cases} \tag{121}$$

It is worth noting that the equation (120) obtained is a nonhomogeneous Fredholm integral equation, and this usually results when converting a nonhomogeneous boundary value problem to its equivalent integral equation. However, homogeneous boundary value problems always lead to homogeneous Fredholm integral equations. Further, we point out here that the solution of boundary value problems is much easier if compared with the solution of its corresponding Fredholm integral equation. This leads to the conclusion that transforming boundary value problem to Fredholm integral equation is less important if compared with transforming initial value problems to Volterra integral equations.

It is of interest to note that the recent developed methods, namely the Adomian decomposition method and the direct computation method, will be introduced in Chapter 2 to handle Fredholm integral Equations. In addition, the traditional methods, namely the successive approximations method and the method of successive substitutions, will be used in Chapter 2 as well.

Exercises 1.6

Derive the equivalent Fredholm integral equation for each of the following boundary value problems:

1. $y'' + 4y = \sin x, 0 < x < 1, \quad y(0) = y(1) = 0$
2. $y'' + 2xy = 1, 0 < x < 1, \quad y(0) = y(1) = 0$
3. $y'' + y = x, 0 < x < 1, \quad y(0) = 1, y(1) = 0$
4. $y'' + y = x, 0 < x < 1, \quad y(0) = 1, y'(1) = 0$

It is to be noted that we will use two developed methods to solve many kinds of integral equations. These two methods mostly give the solution in a series form or in an infinite geometric series. It is therefor useful to present brief summaries of both series. More explanations of these two series can be found in calculus and algebra text books.

1.7   Taylor Series

In this section we will introduce a brief idea on Taylor series. Recall that the Taylor series exists for analytic functions only.

Let $f(x)$ be a function that is infinitely differentiable in an interval $[b, c]$ that contains an interior point a. The Taylor series of $f(x)$ generated at $x = a$ is given by the sigma notation

$$f(x) = \sum_{n=0}^{\infty} \frac{f^{(n)}(a)}{n!}(x - a)^n, \tag{122}$$

which can be written as

$$
\begin{aligned}
f(x) &= f(a) + \frac{f'(a)}{1!}(x-a) + \frac{f''(a)}{2!}(x-a)^2 + \frac{f'''(a)}{3!}(x-a)^3 + \cdots \\
&+ \frac{f^{(n)}(a)}{n!}(x-a)^n + \cdots .
\end{aligned}
\tag{123}
$$

The Taylor series of the function $f(x)$ at $a = 0$ is given by

$$
f(x) = \sum_{n=0}^{\infty} \frac{f^{(n)}(0)}{n!} x^n ,
\tag{124}
$$

which can be written as

$$
f(x) = f(0) + \frac{f'(0)}{1!}x + \frac{f''(0)}{2!}x^2 + \frac{f'''(0)}{3!}x^3 + \cdots + \frac{f^{(n)}(0)}{n!}x^n + \cdots .
\tag{125}
$$

In what follows, we discuss few examples for the derivation of the Taylor series at $x = 0$.

Example 1. Find the Taylor series generated by $f(x) = \cos x$ at $x = 0$.

We list the exponential function and its derivatives as follows:

$f^{(n)}(x)$	$f^{(n)}(0)$
$f(x) = \cos x$	$f(0) = 1,$
$f'(x) = -\sin x$	$f'(0) = 0,$
$f''(x) = -\cos x$	$f''(0) = -1,$
$f'''(x) = \sin x$	$f'''(0) = 0,$
$f^{(iv)}(x) = \cos x$	$f^{(iv)}(0) = 1,$
$\vdots$	

and so on. This gives the Taylor series for $\cos x$ by

$$
\cos x = 1 - \frac{x^2}{2!} + \frac{x^4}{4!} + \cdots ,
\tag{126}
$$

and in a compact form by

$$
\cos x = \sum_{n=0}^{\infty} \frac{(-1)^n}{(2n)!} x^{2n}, \text{ for all } x.
\tag{127}
$$

Similarly, we can easily show that

$$
\cosh x = 1 + \frac{x^2}{2!} + \frac{x^4}{4!} + \cdots ,
\tag{128}
$$

and in a compact form by

$$\cosh x = \sum_{n=0}^{\infty} \frac{1}{(2n)!} x^{2n}, \text{ for all } x. \tag{129}$$

Example 2. Find the Taylor series generated by $f(x) = \sin x$ at $x = 0$.

Proceeding as before we find

$$
\begin{array}{ll}
\underline{f^{(n)}(x)} & \underline{f^{(n)}(0)} \\
f(x) = \sin x & f(0) = 0, \\
f'(x) = \cos x & f'(0) = 1, \\
f''(x) = -\sin x & f''(0) = 0, \\
f'''(x) = -\cos x & f'''(0) = -1, \\
f^{(iv)}(x) = \sin x & f^{(iv)}(0) = 0, \\
f^{(v)}(x) = \cos x & f''(0) = 1, \\
\vdots &
\end{array}
$$

and so on. This gives the Taylor series for $\sin x$ by

$$\sin x = 1 - \frac{x^3}{3!} + \frac{x^5}{5!} + \cdots, \tag{130}$$

and in a compact form by

$$\sin x = \sum_{n=0}^{\infty} \frac{(-1)^n}{(2n+1)!} x^{2n+1}, \text{ for all } x. \tag{131}$$

In a similar way we can show

$$\sinh x = 1 + \frac{x^3}{3!} + \frac{x^5}{5!} + \cdots, \tag{132}$$

and in a compact form by

$$\sin x = \sum_{n=0}^{\infty} \frac{1}{(2n+1)!} x^{2n+1}, \text{ for all } x. \tag{133}$$

In Appendix C, the Taylor series for many well known functions generated at $x = 0$ are given.

It was stated before that the proposed methods that will be used will give the solution in a series form. The obtained series form may converge to a closed form solution, if such a solution exists. Otherwise the truncated series solution may be used for numerical purposes. To get an exact solution, it is normal that we practise for the determination of this solution using Taylor series. In what follows, we study three examples only, and most of the integral equations will lead to these series.

Example 3. Find the closed form function for the following series

$$f(x) = 1 + 3x + \frac{9}{2}x^2 + \frac{9}{2}x^3 + \frac{27}{8}x^4 + \cdots. \qquad (134)$$

This series can be rewritten in the form

$$f(x) = 1 + 3x + \frac{(3x)^2}{2!} + \frac{(3x)^3}{3!} + \frac{(3x)^4}{4!} + \cdots, \qquad (135)$$

that will converge to the exact form

$$f(x) = e^{3x}. \qquad (136)$$

Example 4. Find the closed form function for the following series

$$f(x) = 1 - x - \frac{1}{2}x^2 + \frac{1}{6}x^3 + \frac{1}{24}x^4 - \frac{1}{120}x^5 + \cdots. \qquad (137)$$

We group the series in the form

$$f(x) = (1 - \frac{1}{2!}x^2 + \frac{1}{4!}x^4 + \cdots) - (x - \frac{1}{3!}x^3 + \frac{1}{5!}x^5 + \cdots), \qquad (138)$$

that converges to

$$f(x) = \cos x - \sin x. \qquad (139)$$

Exercises 1.7

Find the closed form function for the following Taylor series:

1. $1 + 2x + 2x^2 + \frac{4}{3}x^3 + \frac{2}{3}x^4 + \cdots$

2. $1 - 3x + \frac{9}{2}x^2 - \frac{9}{2}x^3 + \frac{27}{8}x^4 + \cdots$

3. $x + \frac{1}{2!}x^2 + \frac{1}{3!}x^3 + \frac{1}{4!}x^4 + \cdots$

4. $1 - 2x^2 + \frac{2}{3}x^4 - \frac{4}{45}x^6 + \cdots$

5. $3x - \frac{9}{2}x^3 + \frac{81}{40}x^5 - \frac{243}{560}x^7 + \cdots$

6. $2x + \frac{4}{3}x^3 + \frac{4}{15}x^5 + \frac{8}{315}x^7 + \cdots$

7. $1 + 2x^2 + \dfrac{2}{3}x^4 + \dfrac{4}{45}x^6 + \cdots$

8. $\frac{9}{2}x^2 + \frac{27}{8}x^4 + \frac{81}{80}x^6 + \cdots$

9. $2 - 2x^2 + \dfrac{2}{3}x^4 - \dfrac{4}{45}x^6 + \cdots$

10. $1 + x - \dfrac{1}{6}x^3 + \dfrac{1}{120}x^5 - \dfrac{1}{5040}x^7 + \cdots$

1.8   Infinite Geometric Series

A *geometric series* is a series with a constant ratio between successive terms. The standard form of an infinite geometric series is given by

$$S_n = \sum_{k=0}^{n} a_1 r^k = a_1 + a_1 r + a_1 r^2 + a_1 r^3 + a_1 r^4 + \cdots + a_1 r^n. \qquad (140)$$

An *infinite geometric series* converges if and only if $|r| < 1$, otherwise it diverges. The sum of infinite geometric series, for $|r| < 1$, is given by

$$S_n = \frac{a_1}{1 - r}. \qquad (141)$$

As stated earlier, some of the proposed methods that will be used in this text give the solution as an infinite geometric series. To determine the exact solution in a closed form, it is normal to find the sum of this series. For this reason we will study examples of infinite geometric series.

Example 1. Find the sum of the infinite geometric series

$$1 + \frac{3}{5} + \frac{9}{25} + \frac{27}{125} + \cdots. \qquad (142)$$

It is obvious that the first term is $a_1 = 1$ and the common ratio is $r = \frac{3}{5}$. The sum is therefore given by

$$S = \frac{1}{1 - \frac{3}{5}} = \frac{5}{2}. \qquad (143)$$

Example 2. Find the sum of the infinite geometric series

$$1 - \frac{1}{3} + \frac{1}{9} - \frac{1}{17} + \cdots. \qquad (144)$$

It is obvious that $a_1 = 1$ and $r = -\frac{1}{3}, |r| < 1$. The sum is therefore given by

$$S = \frac{1}{1 + \frac{1}{3}} = \frac{3}{4}. \qquad (145)$$

Example 3. Simplify the following expression

$$S = x^3 + \frac{x}{5} + \frac{x}{15} + \frac{x}{45} + \cdots. \tag{146}$$

The given expression can be rewritten as

$$S = x^3 + \frac{x}{5}\left(1 + \frac{1}{3} + \frac{1}{9} + \cdots\right). \tag{147}$$

It is obvious that the second part is an infinite geometric series, with $a_1 = 1$ and $r = \frac{1}{3}, |r| < 1$. The sum is therefore given by

$$S = x^3 + x\frac{1}{1 - \frac{1}{3}} = x^3 + \frac{3}{2}x. \tag{148}$$

Example 4. Simplify the following expression

$$S = 1 + \frac{\pi}{8}\sec^2 x + \frac{\pi}{16}\sec^2 x + \frac{\pi}{32}\sec^2 x + \cdots. \tag{149}$$

The given expression can be rewritten as

$$S = 1 + \frac{\pi}{8}\sec^2 x \left(1 + \frac{1}{2} + \frac{1}{4} + \cdots\right). \tag{150}$$

It is obvious that the second part is an infinite geometric series, with $a_1 = 1$ and $r = \frac{1}{2}, |r| < 1$. The sum is therefore given by

$$S = 1 + \frac{\pi}{4}\sec^2 x. \tag{151}$$

Example 5. Simplify the following expression

$$S = x^4 + \frac{1}{14}x^2 + \frac{11}{70}x^2 + \frac{11}{350}x^2 + \frac{11}{1750}x^2 + \cdots. \tag{152}$$

The given expression can be rewritten as

$$S = x^4 + \frac{1}{14}x^2 + \frac{11}{70}x^2 \left(1 + \frac{1}{5} + \frac{1}{25} + \cdots\right). \tag{153}$$

It is obvious that the third part is an infinite geometric series, with $a_1 = 1$ and $r = \frac{1}{5}, |r| < 1$. The sum is therefore given by

$$S = x^4 + \frac{1}{14}x^2 + \frac{11}{56}x^2 = x^4 + \frac{15}{56}x^2. \tag{154}$$

Chapter 2

Fredholm Integral Equations

2.1 Introduction

In this chapter we shall be concerned with the nonhomogeneous Fredholm integral equations of the second kind of the form

$$u(x) = f(x) + \lambda \int_a^b K(x,t)u(t)dt, \ \ a \le x \le b, \tag{1}$$

where $K(x,t)$ is the kernel of the integral equation, and λ is a parameter. A considerable amount of discussion will be directed towards the various methods and techniques that are used for solving this type of equations starting with the most recent methods that proved to be highly reliable and accurate. To do this we will naturally focus our study on the *degenerate* or *separable* kernels all through this chapter. The standard form of the *degenerate* or *separable* kernel is given by

$$K(x,t) = \sum_{k=1}^n g_k(x)\, h_k(t). \tag{2}$$

The expressions $x-t$, $x+t$, xt, $x^2 - 3xt + t^2$, etc. are examples of separable kernels. For other well-behaved non-separable kernels, we can convert it to separable in the form (2) simply by expanding these kernels using Taylor's expansion.

Moreover, the kernel $K(x,t)$ is defined to be square integrable in both x and t in the square $a \le x \le b$, $a \le t \le b$ if the following *regularity*

35

condition

$$\int_a^b \int_a^b K(x,t)\, dx\, dt < \infty, \tag{3}$$

is satisfied. This condition gives rise to the development of the solution of the Fredholm integral equation (1). It is also convenient to state, without proof, the so-called *Fredholm Alternative Theorem* that relates the solutions of homogeneous and nonhomogeneous Fredholm integral equations.

Fredholm Alternative Theorem

The nonhomogeneous Fredholm integral equation (1) has one and only one solution if the only solution to the homogeneous Fredholm integral equation

$$u(x) = \lambda \int_a^b K(x,t)u(t)dt, \tag{4}$$

is the trivial solution $u(x) = 0$.

We end this section by introducing the necessary condition that will guarantee a unique solution to the integral equation (1) in the interval of discussion. Considering (2), if the kernel $K(x,t)$ is real, continuous and bounded in the square $a \le x \le b$ and $a \le t \le b$, i.e. if

$$|K(x,t)| \le M, \quad a \le x \le b, \quad \text{and} \quad a \le t \le b, \tag{5}$$

and if $f(x) \ne 0$, and continuous in $a \le x \le b$, then the necessary condition that will guarantee that (1) has only a unique solution is given by

$$|\lambda|\, M\, (b-a) < 1. \tag{6}$$

It is important to note that a continuous solution to Fredholm integral equation may exist, even though the condition (6) is not satisfied. This may be clearly seen by considering the equation

$$u(x) = -4 + \int_0^1 (2x + 3t)u(t)dt. \tag{7}$$

In this example, $\lambda = 1, |K(x,t)| \le 5$ and $(b-a) = 1$; therefore

$$|\lambda|\, M\, (b-a) = 5 \not< 1. \tag{8}$$

Accordingly, the necessary condition (6) fails to hold, but in fact the integral equation (7) has an exact solution given by

$$u(x) = 4x, \tag{9}$$

and this can be justified through direct substitution.

As indicated in our objective for a first course in integral equations, we will pay more attention to the practical techniques for solving integral equations rather than the abstract theorems. In the following we will discuss several methods that handle successfully the Fredholm integral equations of the second kind starting with the most recent methods as indicated earlier.

2.2 The Adomian Decomposition Method

Adomian [1] developed the so-called Adomian decomposition method or simply the *decomposition method* (ADM). The method was well introduced by Adomian in his recent books [1] and [2]. The method proved to be reliable and effective for a wide class of equations, differential and integral equations, linear and nonlinear models. The method provides the solution in a series form as will be seen later. The method was applied mostly to ordinary and partial differential equations, and was rarely used for integral equations in [1] and [2]. The concept of convergence of the solution obtained by this method was addressed extensively in the literature. The convergence concept is beyond the scope of this text. However, the decomposition method can be successfully applied towards linear and nonlinear integral equations.

In the decomposition method we usually express the solution $u(x)$ of the integral equation (1) in a series form defined by

$$u(x) = \sum_{n=0}^{\infty} u_n(x). \tag{10}$$

Substituting the decomposition (10) into both sides of (1) yields

$$\sum_{n=0}^{\infty} u_n(x) = f(x) + \lambda \int_a^b K(x,t) \left(\sum_{n=0}^{\infty} u_n(t) \right) dt, \tag{11}$$

or equivalently

$$\begin{aligned} u_0(x) + u_1(x) + u_2(x) + \cdots \ &= f(x) \ + \lambda \int_a^b K(x,t) u_0(t) dt \\ &+ \lambda \int_a^b K(x,t)\, u_1(t) dt \\ &+ \lambda \int_a^b K(x,t)\, u_2(t) dt \\ &+ \cdots \end{aligned} \tag{12}$$

The components $u_0(x), u_1(x), u_2(x), u_3(x), \ldots$ of the unknown function $u(x)$ are completely determined in a recurrent manner if we set

$$u_0(x) = f(x), \tag{13}$$

$$u_1(x) = \lambda \int_a^b K(x,t) u_0(t) dt, \tag{14}$$

$$u_2(x) = \lambda \int_a^b K(x,t)u_1(t)dt, \tag{15}$$

$$u_3(x) = \lambda \int_a^b K(x,t)u_2(t)dt, \tag{16}$$

and so on. The above discussed scheme for the determination of the components $u_0(x), u_1(x), u_2(x), u_3(x), ...$ of the solution $u(x)$ of Eq. (1) can be written in a recursive manner by

$$u_0(x) = f(x), \tag{17}$$

$$u_{n+1}(x) = \lambda \int_a^b K(x,t)u_n(t)dt, \; n \geq 0. \tag{18}$$

In view of (17) and (18), the components $u_0(x), u_1(x), u_2(x), u_3(x), ...$ follow immediately. With these components determined, the solution $u(x)$ of (1) is readily determined in a series form using the decomposition (10). It is important to note that the obtained series for $u(x)$ converges to the exact solution in a closed form if such a solution exists as will be seen later. However, for concrete problems, where exact solution cannot be evaluated, a truncated series $\sum_{n=0}^{k} u_n(x)$ is usually used to approximate the solution $u(x)$ and this can be used for numerical purposes. We point out here that few terms of the truncated series usually provide the higher accuracy level of the approximate solution if compared with the existing numerical techniques. The decomposition technique proved to be effective and reliable even if applied to nonlinear Fredholm integral equations as will be discussed in a forthcoming chapter.

In the following we discuss some examples that illustrate the decomposition method outlined above.

Example 1. We first consider the Fredholm integral equation of the second kind

$$u(x) = \frac{9}{10}x^2 + \int_0^1 \frac{1}{2} x^2 t^2 u(t) \, dt. \tag{19}$$

It is clear that $f(x) = \frac{9}{10}x^2, \lambda = 1, K(x,t) = \frac{1}{2}x^2t^2$. To evaluate the components $u_0(x), u_1(x), u_2(x), ...$ of the series solution, we use the recursive scheme (17) and (18) to find

$$u_0(x) = \frac{9}{10}x^2, \tag{20}$$

$$u_1(x) = \int_0^1 \frac{1}{2} x^2 t^2 u_0(t)dt,$$

$$= \int_0^1 \frac{1}{2} x^2 \frac{9}{10} t^4 dt$$

$$= \frac{9}{100} x^2, \tag{21}$$

$$u_2(x) = \int_0^1 \frac{1}{2} x^2 t^2 u_1(t) dt$$

$$= \int_0^1 \frac{1}{2} x^2 t^2 \frac{9}{100} t^2 dt$$

$$= \frac{9}{1000} x^2, \tag{22}$$

and so on. Noting that

$$u(x) = u_0(x) + u_1(x) + u_2(x) + \cdots, \tag{23}$$

we can easily obtain the solution in a series form given by

$$u(x) = \frac{9}{10} x^2 + \frac{9}{100} x^2 + \frac{9}{1000} x^2 + \cdots, \tag{24}$$

so that the solution of (19) in a closed form

$$u(x) = x^2, \tag{25}$$

follows immediately upon using the formula for the sum of the infinite geometric series.

Example 2. We next consider the Fredholm integral equation

$$u(x) = \cos x + 2x + \int_0^\pi xt u(t) \, dt. \tag{26}$$

Proceeding as in Example 1, we set

$$u_0(x) = \cos x + 2x, \tag{27}$$

$$u_1(x) = \int_0^\pi xt u_0(t) dt$$

$$= \int_0^\pi xt \left(\cos t + 2t \right) dt$$

$$= \left(-2 + \frac{2}{3} \pi^3 \right) x, \tag{28}$$

$$u_2(x) = \int_0^\pi xt u_1(t) dt$$

$$= \int_0^\pi x \left(-2 + \frac{2}{3}\pi^3 \right) t^2 dt$$

$$= \left(-\frac{2}{3}\pi^3 + \frac{2}{9}\pi^6 \right) x. \tag{29}$$

Consequently, the solution of (26) in a series form is given by

$$u(x) = \cos x + 2x + \left(-2 + \frac{2}{3}\pi^3 \right) x + \left(-\frac{2}{3}\pi^3 + \frac{2}{9}\pi^6 \right) x + \cdots \tag{30}$$

and in a closed form

$$u(x) = \cos x, \tag{31}$$

by eliminating the so-called self-cancelling noise terms between various components of $u(x)$. The answer obtained can be justified through substitution. The self-cancelling noise terms are defined to be similar terms with opposite signs that will vanish in the limit. The phenomenon of noise terms will be presented in a forthcoming section.

Example 3. We consider here the Fredholm integral equation

$$u(x) = e^x - 1 + \int_0^1 t\,u(t)\,dt. \tag{32}$$

Applying the decomposition technique as discussed before we find

$$u_0(x) = e^x - 1, \tag{33}$$

$$u_1(x) = \int_0^1 tu_0(t)dt$$

$$= \int_0^1 t\left(e^t - 1\right) dt$$

$$= \frac{1}{2}, \tag{34}$$

$$u_2(x) = \int_0^1 tu_1(t)dt$$

$$= \int_0^1 \frac{1}{2}t\,dt$$

$$= \frac{1}{4}. \tag{35}$$

The determination of the components (33)-(35) yields the solution of the equation (32) in a series form given by

$$u(x) = e^x - 1 + \frac{1}{2}\left(1 + \frac{1}{2} + \frac{1}{4} + \cdots \right), \tag{36}$$

where we can easily obtain the solution in a closed form given by

$$u(x) = e^x, \tag{37}$$

by evaluating the sum of the infinite geometric series in the right hand side of Eq. (36). Recall that the sum of the infinite geometric series was presented in the previous chapter.

It is important to note that the evaluation of the components $u_0(x)$, $u_1(x), u_2(x), \ldots$ is simple as we observed from the examples above. However, we can still reduce the size of calculations by using a modified version of the decomposition method. In this modified approach, we often need to evaluate the first two components $u_0(x)$ and $u_1(x)$ only. In what follows, we introduce the modified decomposition method suggested by Wazwaz [59].

2.2.1   The Modified Decomposition Method

It is worth noting that the Adomian decomposition method may be sometimes implemented in a different but easier manner in order to facilitate the computational work. It is recommended to apply the modified decomposition method, developed by Wazwaz [59], for cases where the nonhomogeneous part $f(x)$ in (1) consists of a combination of many terms. This modified technique, as will be seen later, will minimize the volume of calculations and reduce the several integral evaluations that result in applying the standard Adomian decomposition method.

It is also of interest, before giving a clear discussion of this method, to note that this modified technique will be carried out with promising results in Volterra integral equations and nonlinear integral equations in forthcoming chapters. The technique avoids the cumbersome integrations of other methods.

In the modified method, we split the given function $f(x)$ into two parts defined by

$$f(x) = f_0(x) + f_1(x), \tag{38}$$

where $f_0(x)$ consists of number of terms of $f(x)$, and $f_1(x)$ includes the remaining terms of $f(x)$. We note that a necessary condition is required to apply this approach in that $f(x)$ should consist of more than one term as shown by (38). In view of (38), the integral equation (1) becomes

$$u(x) = f_0(x) + f_1(x) + \lambda \int_a^b K(x,t)u(t)dt, \ \ a \le x \le b. \tag{39}$$

Substituting the decomposition series (10) into both sides of (39), and using few terms of the expansion we obtain

$$u_0(x) + u_1(x) + u_2(x) + \cdots \; = f_0(x) + f_1(x) \;\; + \lambda \int_a^b K(x,t)u_0(t)dt$$

$$+ \lambda \int_a^b K(x,t)\, u_1(t)dt$$

$$+ \lambda \int_a^b K(x,t)\, u_2(t)dt$$

$$+ \cdots.$$

$$(40)$$

The components $u_0(x), u_1(x), u_2(x), u_3(x), \ldots$ of the unknown function $u(x)$ can be completely determined in a recurrent manner if we assign $f_0(x)$ only to the zeroth component $u_0(x)$, whereas the function $f_1(x)$ will be added to the formula of the component $u_1(x)$ given before in Eq. (14). In other words the modified decomposition method works elegantly if we set

$$u_0(x) \;\; = \;\; f_0(x), \tag{41}$$

$$u_1(x) \;\; = \;\; f_1(x) + \lambda \int_a^b K(x,t)u_0(t)dt, \tag{42}$$

$$u_2(x) \;\; = \;\; \lambda \int_a^b K(x,t)u_1(t)dt, \tag{43}$$

$$u_3(x) \;\; = \;\; \lambda \int_a^b K(x,t)u_2(t)dt, \tag{44}$$

and so on. The above discussed scheme for the determination of the components $u_0(x), u_1(x), u_2(x), u_3(x), \ldots$ of the solution $u(x)$ of the equation (1) can be written in a modified recursive manner by

$$u_0(x) \;\; = \;\; f_0(x), \tag{45}$$

$$u_1(x) \;\; = \;\; f_1(x) + \lambda \int_a^b K(x,t)u_0(t)dt, \tag{46}$$

$$u_{n+1}(x) \;\; = \;\; \lambda \int_a^b K(x,t)u_n(t)dt, \; n \geq 1. \tag{47}$$

Recall that in most problems we need to use two iterations only where we need to use (45) and (46).

The modified decomposition scheme can be explained by the following illustrative examples:

Example 4. We consider here the Fredholm integral equation

$$u(x) = e^{3x} - \frac{1}{9}\left(2e^3 + 1\right)x + \int_0^1 xt\,u(t)\,dt. \tag{48}$$

To apply the modified decomposition scheme as discussed above, we first split the function $f(x)$ into

$$f_0(x) = e^{3x}, \tag{49}$$

and

$$f_1(x) = -\frac{1}{9}\left(2e^3 + 1\right)x. \tag{50}$$

Therefore, we set

$$u_0(x) = e^{3x}, \tag{51}$$

and

$$\begin{aligned}
u_1(x) &= -\tfrac{1}{9}\left(2e^3 + 1\right)x + \int_0^1 xt u_0(t)dt \\
&= -\tfrac{1}{9}\left(2e^3 + 1\right)x + x\int_0^1 te^{3t}dt \\
&= 0.
\end{aligned} \tag{52}$$

In view of (52), we conclude that $u_n = 0$, $n \geq 1$. The exact solution

$$u(x) = e^{3x}, \tag{53}$$

follows immediately.

Example 5. We consider here the Fredholm integral equation

$$u(x) = \sin^{-1} x + \left(\frac{\pi}{2} - 1\right)x - \int_0^1 x\,u(t)\,dt. \tag{54}$$

Applying the modified decomposition method as discussed above, we first split the function $f(x)$ into

$$f_0(x) = \sin^{-1} x, \tag{55}$$

and

$$f_1(x) = \left(\frac{\pi}{2} - 1\right)x. \tag{56}$$

Therefore, we set

$$
\begin{aligned}
u_0(x) &= \sin^{-1} x, && (57)\\
u_1(x) &= \left(\frac{\pi}{2} - 1\right) x - \int_0^1 x u_0(t)\,dt\\
&= \left(\frac{\pi}{2} - 1\right) x - x \int_0^1 \sin^{-1} t\,dt\\
&= 0. && (58)
\end{aligned}
$$

Consequently, the components $u_n(x) = 0$, $n \geq 1$. The exact solution

$$
u(x) = \sin^{-1} x, \tag{59}
$$

is readily obtained.

Example 6. We consider here the Fredholm integral equation

$$
u(x) = \sin x + \cos x - 2x + \frac{\pi}{2} + \int_0^{\frac{\pi}{2}} (x - t)\,u(t)\,dt. \tag{60}
$$

We first split the function $f(x)$ into

$$
\begin{aligned}
f_0(x) &= \sin x + \cos x,\\
f_1(x) &= -2x + \frac{\pi}{2}.
\end{aligned} \tag{61}
$$

We then set

$$
\begin{aligned}
u_0(x) &= \sin x + \cos x,\\
u_1(x) &= -2x + \frac{\pi}{2} + \int_0^{\frac{\pi}{2}} (x - t)\,u_0(t)\,dt = 0.
\end{aligned} \tag{62}
$$

Consequently, the components $u_n(x) = 0$, $n \geq 1$. The exact solution

$$
u(x) = \sin x + \cos x, \tag{63}
$$

is readily obtained.

Example 7. We finally consider the Fredholm integral equation

$$
u(x) = 1 + x + \sec^2 x - 32x^2 - 8\pi x^2 - \pi^2 x^2 + \int_0^{\frac{\pi}{4}} 32x^2\, u(t)\,dt. \tag{64}
$$

We divide the function $f(x)$ into

$$
\begin{aligned}
f_0(x) &= 1 + x + \sec^2,\\
f_1(x) &= -32x^2 - 8\pi x^2 - \pi^2 x^2.
\end{aligned} \tag{65}
$$

We then set

$$\begin{aligned}
u_0(x) &= 1 + x + \sec^2, \\
u_1(x) &= -32x^2 - 8\pi x^2 - \pi^2 x^2 + \int_0^{\frac{\pi}{4}} 32x^2 \, u_0(t) \, dt = 0.
\end{aligned} \tag{66}$$

Consequently, the components $u_n(x) = 0$, $n \geq 1$. The exact solution

$$u(x) = 1 + x + \sec^2, \tag{67}$$

is readily obtained.

This confirms our belief that the decomposition method and the modified decomposition method introduce the solution of Fredholm integral equation in the form of a rapidly convergent power series with elegantly computable terms. However, if $f(x)$ consists of more than one term, the modified decomposition method minimizes the volume of the computational work.

2.2.2 The Noise Terms Phenomenon

In [62], Wazwaz examined the noise terms phenomenon which accelerates the convergence of the Adomian decomposition method. The noise terms phenomenon demonstrates a useful tool for fast convergence of the solution. In [62], it was proved that the noise terms phenomenon may appear only for nonhomogeneous PDEs of any order or nonhomogeneous integral equations of any kind. The noise terms, if exist, it will appear in all components. The noise terms that may exist in the components u_0 and u_1, will give the solution in a closed form by using the components $u_0(x)$ and $u_1(x)$.

The noise terms are the identical terms with opposite signs that arise in the components, and may exist only for nonhomogeneous equations. It was found that by canceling the noise terms between $u_0(x)$ and $u_1(x)$, even though u_1 contains other terms, the remaining non-canceled terms of $u_0(x)$ may give the exact solution of the equation. Notice that it is necessary to verify that the non-canceled terms of $u_0(x)$ satisfy the PDE or the integral equation. In case the non-canceled terms of $u_0(x)$ did not satisfy the given equation, or the noise terms did not appear, then we proceed to determine more components of the series solution.

It was shown by many authors that non homogeneity condition is not sufficient to give noise terms. Moreover, it is necessary that the zeroth component $u_0(x)$ should contain the exact solution $u(x)$ among other terms, and this only may give noise terms about components. Moreover, noise terms may appear if the exact solution $u(x)$ is part of the zeroth component $u_0(x)$. The noise terms phenomenon will be explained by studying the following illustrative examples.

Example 8. Solve the Fredholm integral equation by using the noise terms phenomenon

$$u(x) = x \cos x + 2x + \int_0^\pi x u(t)\, dt. \tag{68}$$

Using the Adomian method, we set the recurrence relation

$$
\begin{aligned}
u_0(x) &= x \cos x + 2x, \\
u_{k+1}(x) &= \int_0^\pi x u_k(t)\, dt, \ k \geq 0.
\end{aligned}
\tag{69}
$$

This gives

$$
\begin{aligned}
u_0(x) &= x \cos x + 2x, \\
u_1(x) &= \int_0^\pi x u_0(t)\, dt = -2x + \pi^2 x.
\end{aligned}
\tag{70}
$$

The noise terms $\pm 2x$ appear in $u_0(x)$ and $u_1(x)$. Canceling this term from the zeroth component $u_0(x)$ gives the exact solution

$$u(x) = x \cos x, \tag{71}$$

that justifies the integral equation. The other terms of $u_1(x)$ vanish in the limit with other terms of the other components.

Example 9. Solve the Fredholm integral equation by using the noise terms phenomenon

$$u(x) = \cos x - \sin x + 2x - \frac{\pi}{2}x + \int_0^{\frac{\pi}{2}} x t u(t)\, dt. \tag{72}$$

The standard Adomian method gives the recurrence relation

$$
\begin{aligned}
u_0(x) &= \cos x - \sin x + 2x - \frac{\pi}{2}x, \\
u_{k+1}(x) &= \int_0^{\frac{\pi}{2}} x t u_k(t)\, dt, \ k \geq 0.
\end{aligned}
\tag{73}
$$

This gives

$$
\begin{aligned}
u_0(x) &= \cos x - \sin x + 2x - \frac{\pi}{2}x, \\
u_1(x) &= \int_0^{\frac{\pi}{2}} x t u_0(t)\, dt = -2x + \frac{\pi}{2}x + \frac{\pi^3}{12}x - \frac{\pi^4}{48}x.
\end{aligned}
\tag{74}
$$

The noise terms $\mp \frac{\pi}{2}x$ and $\pm 2x$ appear in $u_0(x)$ and $u_1(x)$. Canceling these terms from $u_0(x)$ gives the exact solution

$$u(x) = \cos x - \sin x, \tag{75}$$

that satisfies the integral equation.

Example 10. Solve the Fredholm integral equation by using the noise terms phenomenon

$$u(x) = \frac{\pi}{2}x - x + x\tan^{-1}x - \int_{-1}^{1} xu(t)\, dt. \tag{76}$$

We use the recurrence relation

$$
\begin{aligned}
u_0(x) &= \frac{\pi}{2}x - x + x\tan^{-1}x, \\
u_{k+1}(x) &= -\int_{-1}^{1} xu_k(t)\, dt, \ k \ge 0.
\end{aligned}
\tag{77}
$$

This gives

$$
\begin{aligned}
u_0(x) &= \frac{\pi}{2}x - x + x\tan^{-1}x, \\
u_1(x) &= -\int_{-1}^{1} xu_0(t)\, dt = -\frac{\pi}{2}x + x.
\end{aligned}
\tag{78}
$$

Canceling these terms from $u_0(x)$ gives the exact solution

$$u(x) = x\tan^{-1}x, \tag{79}$$

that satisfies the integral equation.

Exercises 2.2

In exercises 1-12, solve the following Fredholm integral equations by using the *Adomian decomposition method*

1. $u(x) = \frac{13}{3}x - \frac{1}{4}\int_{0}^{1} xtu(t)dt.$

2. $u(x) = x^3 - \frac{1}{5}x + \int_{0}^{1} xtu(t)dt.$

3. $u(x) = x^2 + \int_{0}^{1} xtu(t)dt.$

4. $u(x) = e^x + e^{-1}\int_{0}^{1} u(t)dt.$

5. $u(x) = x + \sin x - x\int_{0}^{\pi/2} u(t)dt.$

6. $u(x) = x + \cos x - 2x\int_{0}^{\pi/6} u(t)dt.$

7. $u(x) = \cos(4x) + \frac{1}{4}x - \int_{0}^{\pi/8} xu(t)dt.$

8. $u(x) = \sinh x - e^{-1}x + \int_0^1 xtu(t)dt.$

9. $u(x) = 2e^{2x} + (1 - e^2) x + \int_0^1 xu(t)dt.$

10. $u(x) = 1 + \sec^2 x - \int_0^{\pi/4} u(t)dt.$

11. $u(x) = \sin x + \int_{-1}^1 e^{\sin^{-1} x} u(t)dt.$

12. $u(x) = \tan x - \int_{-\pi/3}^{\pi/3} e^{\tan^{-1} x} u(t)dt.$

In exercises 13-20 solve the given Fredholm integral equations by using the *modified decomposition method*:

13. $u(x) = \tan^{-1} x + \frac{1}{2} \left(\ln 2 - \frac{\pi}{2} \right) x + \int_0^1 xu(t)dt.$

14. $u(x) = \cosh x + (\sinh 1) x + (e^{-1} - 1) - \int_0^1 (x - t)u(t)dt.$

15. $u(x) = \frac{1}{1 + x^2} + 2x \sinh(\pi/4) - x \int_{-1}^1 e^{\tan^{-1} t} u(t)dt.$

16. $u(x) = \frac{1}{\sqrt{1 - x^2}} + \left(e^{\pi/6} - 1 \right) x - x \int_0^{1/2} e^{\sin^{-1} t} u(t)dt.$

17. $u(x) = \frac{1}{1 + x^2} + \frac{\pi^2}{32} x - x \int_0^1 \tan^{-1} tu(t)dt.$

18. $u(x) = \cos^{-1} x - \pi x + \int_{-1}^1 xu(t)dt.$

19. $u(x) = x \tan^{-1} x + \left(\frac{\pi}{4} - \frac{1}{2} \right) x - \int_0^1 xu(t)dt.$

20. $u(x) = x \sin^{-1} x + 1 - \left(\frac{\pi}{8} + 1 \right) x + \int_0^1 xu(t)dt.$
Hint: $f_1(x) = x \sin^{-1} x + 1.$

In exercises 21-26 solve the given Fredholm integral equations by using the *noise terms phenomenon*:

21. $u(x) = \frac{\sin x}{1 + \sin x} + x - \frac{\pi}{2} x + \int_0^{\frac{\pi}{2}} xu(t)dt.$

22. $u(x) = \frac{\sin x}{1 + \cos x} - x \ln(2) + \int_0^{\frac{\pi}{2}} xu(t)dt.$

23. $u(x) = \frac{\sec^2 x}{1 + \tan x} - x \ln(2) + \int_0^{\frac{\pi}{4}} xu(t)dt.$

24. $u(x) = 1 + \sin x - x - \dfrac{\pi^2}{2}x + \displaystyle\int_0^{\frac{\pi}{2}} xtu(t)dt.$

25. $u(x) = 1 + \sin x + \cos x - 2x - \dfrac{\pi}{2}x + \dfrac{\pi}{2} + \dfrac{\pi^2}{8} + \displaystyle\int_0^{\frac{\pi}{2}} (x - t)u(t)dt.$

26. $u(x) = x \sin x - 2 - x + \pi + \displaystyle\int_0^{\frac{\pi}{2}} (x - t)u(t)dt.$

2.3 The Variational Iteration Method

In this section we will present the *variational iteration method* that was used recently in the literature to handle both differential and integral equations, linear and nonlinear. Recall that the Adomian decomposition method gives the components of the decomposition series by using a recurrence relation. Unlike the ADM, the variational iteration method gives successive approximations of the solution that may converge rapidly to the exact solution if such a solution exists. However, for concrete problems the obtained approximations can be used for numerical reasons.

It is interesting to note that the variational iteration method (VIM) is used for a differential equation, ordinary or partial, and for an integro-differential equation. This means that to use this method for solving integral equations, we first should convert the integral equation to its equivalent differential equation, or to its equivalent integro-differential equation, by using any appropriate method. Unlike the Adomian method that can be used directly to solve an integral equation, the VIM will be employed to the converted differential or integro-differential equation.

In this section, we will apply the variational iteration method to handle Fredholm integral equation. The method works effectively if the kernel $K(x,t)$ is separable of the form $K(x,t) = g(x)h(t)$. The Fredholm integral equation can be converted to an identical Fredholm integro-differential equation by differentiating both sides, where an initial condition should also be derived. For simplicity, we will study only the cases where $g(x) = x^n, n \geq 1$. In what follows we will present the main steps for using this method.

The standard Fredholm integral equation is of the form

$$u(x) = f(x) + \int_a^b K(x,t)u(t)dt, \qquad (80)$$

or equivalently

$$u(x) = f(x) + g(x)\int_a^b h(t)u(t)dt, \quad K(x,t) = g(x)h(t). \qquad (81)$$

Recall that the integral at the right side of (81) depends on t only, hence it is equivalent to a constant. Differentiating both sides of (81) with respect to x gives

$$u^{'}(x) = f^{'}(x) + g^{'}(x) \int_a^b h(t)u(t)dt. \qquad (82)$$

The variational iteration method admits the use of a correction functional for the integro-differential equation (82) in the form

$$u_{n+1}(x) = u_n(x) + \int_0^x \lambda(\xi) \left(u_n^{'}(\xi) - f^{'}(\xi) - g^{'}(\xi) \int_a^b h(r)\tilde{u}_n(r)\,dr \right) d\xi, \qquad (83)$$

where λ is a general Lagrange multiplier. Note that the Lagrange multiplier λ may be a constant or a function, and $\tilde{u}_n$ is a restricted value that means it behaves as a constant, hence $\delta\tilde{u}_n = 0$, where δ is the variational derivative. The Lagrange multiplier λ can be identified optimally via the variational theory as proved in the literature.

The variational iteration method depends mainly on two essential steps. We first should determine the Lagrange multiplier $\lambda(\xi)$ that can be identified optimally via the variational theory where integration by parts should be used. As stated before, λ may be a constant or a function, and it is different from one question to another. A list of some of these Lagrange multipliers will be given later, but this list does not cover all differential or integral problems. However for our present use in this chapter, $\lambda(\xi) = -1$ for first order integro-differential equations. Having determined λ, an iteration formula, without restricted variation, given by

$$u_{n+1}(x) = u_n(x) - \int_0^x \left(u_n^{'}(\xi) - f^{'}(\xi) - g^{'}(\xi) \int_a^b h(r)u_n(r)\,dr \right) d\xi, \qquad (84)$$

is used to determine the successive approximations $u_{n+1}(x), n \geq 0$ of the solution $u(x)$. Notice that $u_n(x)$ gives the successive approximations of the solution and not the components as in the case when Adomian method is used. In other words, the correction functional will give several approximations of the solution.

The zeroth approximation u_0 can be any selective function. However, using the given initial value $u(0)$ is preferably used for the selective zeroth approximation u_0 as will be seen later. Consequently, the solution is given by

$$u(x) = \lim_{n \to \infty} u_n(x). \qquad (85)$$

The determination of the Lagrange multiplier is essential for the use of the correction functional. In what follows, we summarize some iteration formulae that show ODE, its corresponding Lagrange multipliers for this kind of ODEs, and its correction functional respectively [18]:

$$(i) \begin{cases} u' + f(u(\xi), u'(\xi)) = 0, \ \lambda = -1, \\ u_{n+1} = u_n - \int_0^x \left[u'_n + f(u_n, u'_n) \right] d\xi, \end{cases}$$

$$(ii) \begin{cases} u'' + f(u(\xi), u'(\xi), u''(\xi)) = 0, \ \lambda = (\xi - x), \\ u_{n+1} = u_n + \int_0^x (\xi - x) \left[u''_n + f(u_n, u'_n, u''_n) \right] d\xi, \end{cases}$$

$$(iii) \begin{cases} u''' + f(u(\xi), u'(\xi), u''(\xi), u'''(\xi)) = 0, \ \lambda = -\frac{1}{2!}(\xi - x)^2, \\ u_{n+1} = u_n - \int_0^x \frac{1}{2!}(\xi - x)^2 \left[u'''_n + f(u_n, ..., u'''_n) \right] d\xi, \end{cases}$$

$$(iv) \begin{cases} u^{(iv)} + f(u(\xi), u'(\xi), u''(\xi), u'''(\xi), u^{(iv)}(\xi)) = 0, \ \lambda = \frac{1}{3!}(\xi - x)^3, \\ u_{n+1} = u_n + \int_0^x \frac{1}{3!}(\xi - x)^3 \left[u'''_n + f(u_n, u'_n, ..., u_n^{(iv)}) \right] d\xi, \end{cases}$$

and generally

$$(v) \begin{cases} u^{(n)} + f(u(\xi), u'(\xi), \cdots, u^{(n)}(\xi)) = 0, \lambda = (-1)^n \frac{1}{(n-1)!}(\xi - x)^{(n-1)}, \\ u_{n+1} = u_n + (-1)^n \int_0^x \frac{1}{(n-1)!}(\xi - x)^{(n-1)} \left[u'''_n + f(u_n, ..., u_n^{(n)}) \right] d\xi, \end{cases}$$

for $n \geq 1$.

The variational iteration method will be illustrated by studying the following Fredholm integral equations.

Example 1. Use the variational iteration method to solve the Fredholm integral equation

$$u(x) = xe^x - x + x \int_0^1 u(t)dt. \tag{86}$$

Differentiating both sides of this equation with respect to x yields

$$u'(x) = xe^x + e^x - 1 + \int_0^1 u(t)dt, u(0) = 0. \tag{87}$$

The correction functional for this equation is given by

$$u_{n+1}(x) = u_n(x) - \int_0^x \left(u'_n(\xi) - \xi e^\xi - e^\xi + 1 - \int_0^1 u_n(r) \, dr \right) d\xi, \tag{88}$$

where we used $\lambda = -1$ for first-order integro-differential equations. It is preferable to select $u_0(x) = u(0) = 1$. Using this selection into the correction functional gives the following successive approximations

$$u_0(x) = 0,$$
$$u_1(x) = u_0 - \int_0^x \left(u_0'(\xi) - \xi e^\xi - e^\xi + 1 - \int_0^1 u_0(r)dr \right) d\xi = xe^x - x,$$
$$u_2(x) = u_1 - \int_0^x \left(u_1'(\xi) - \xi e^\xi - e^\xi + 1 - \int_0^1 u_1(r)dr \right) d\xi = xe^x - \tfrac{1}{2}x,$$
$$u_3(x) = u_2 - \int_0^x \left(u_2'(\xi) - \xi e^\xi - e^\xi + 1 - \int_0^1 u_2(r)dr \right) d\xi = xe^x - \tfrac{1}{4}x,$$
$$\vdots$$
$$u_n(x) = xe^x - \frac{1}{2^{n-1}}x, n \geq 1.$$

$$(89)$$

The VIM admits the use of

$$u(x) = \lim_{n \to \infty} u_n(x) = xe^x. \tag{90}$$

Example 2. Use the variational iteration method to solve the Fredholm integral equation

$$u(x) = \cos x - x + x \int_0^{\frac{\pi}{2}} u(t)dt. \tag{91}$$

Differentiating both sides of this equation with respect to x gives

$$u'(x) = -\sin x - 1 + \int_0^{\frac{\pi}{2}} u(t)dt, u(0) = 1. \tag{92}$$

The correction functional for this equation is given by

$$u_{n+1}(x) = u_n(x) - \int_0^x \left(u_n'(\xi) + \sin \xi + 1 - \int_0^{\frac{\pi}{2}} u_n(r)\, dr \right) d\xi, \tag{93}$$

where we used $\lambda = -1$ for first-order integro-differential equations. The initial condition $u(0) = 1$ is obtained by substituting $x = 0$ into (91).

We can use the initial condition to select $u_0(x) = u(0) = 0$. Using this selection into the correction functional gives the following successive approximations

$$u_0(x) = 1,$$
$$u_1(x) = u_0(x) - \int_0^x \left(u_0'(\xi) + \sin \xi + 1 - \int_0^{\frac{\pi}{2}} u_0(r)\, dr \right) d\xi$$
$$= \cos x - x + \tfrac{\pi}{2}x,$$
$$u_2(x) = u_1(x) - \int_0^x \left(u_1'(\xi) + \sin \xi + 1 - \int_0^{\frac{\pi}{2}} u_1(r)\, dr \right) d\xi$$
$$= (\cos x - x) + (x - \tfrac{\pi^2}{8}x) + \tfrac{\pi^3}{16}x,$$
$$u_3(x) = u_2(x) - \int_0^x \left(u_2'(\xi) + \sin \xi + 1 - \int_0^{\frac{\pi}{2}} u_2(r)\, dr \right) d\xi$$
$$= (\cos x - x) + (x - \tfrac{\pi^2}{8}x) + (\tfrac{\pi^2}{8}x - \tfrac{\pi^3}{16}x) + (\tfrac{\pi^3}{16}x - \cdots) + \cdots,$$

$$(94)$$

and so on. Canceling the noise terms, the exact solution is given by

$$u(x) = \cos x. \tag{95}$$

Example 3. Use the variational iteration method to solve the Fredholm integral equation

$$u(x) = x^2 - \frac{1}{3}x + \int_0^1 xu(t)dt. \tag{96}$$

Differentiating both sides of this equation with respect to x gives

$$u'(x) = 2x - \frac{1}{3} + \int_0^1 u(t)dt, u(0) = 0. \tag{97}$$

The correction functional for this equation is given by

$$u_{n+1}(x) = u_n(x) - \int_0^x \left(u_n'(\xi) - 2\xi + \frac{1}{3} - \int_0^1 u_n(r)\,dr \right) d\xi. \tag{98}$$

The initial condition $u(0) = 0$ is obtained by substituting $x = 0$ into (96). Using this selection into the correction functional gives the following successive approximations

$$
\begin{aligned}
u_0(x) &= 0, \\
u_1(x) &= x^2 - \frac{1}{3}x, \\
u_2(x) &= x^2 - \frac{1}{6}x, \\
u_3(x) &= x^2 - \frac{1}{12}x, \\
&= \vdots,
\end{aligned}
\tag{99}
$$

and so on. This in turn gives

$$u_n(x) = x^2 - \frac{1}{3}(\frac{1}{2})^{n-1}x, n \geq 1. \tag{100}$$

This converges to the exact solution

$$u(x) = x^2. \tag{101}$$

Exercises 2.3

Solve the following Fredholm integral equations by using the *variational iteration method*

1. $u(x) = x^3 - \frac{1}{5}x + \int_0^1 xtu(t)dt$

2. $u(x) = e^x - x + \int_0^1 xtu(t)dt$

3. $u(x) = \dfrac{2}{3}x + \displaystyle\int_0^1 xtu(t)dt$

4. $u(x) = x^2 + x^4 - \dfrac{5}{12}x + \displaystyle\int_0^1 xtu(t)dt$

5. $u(x) = e^x + 2x - \dfrac{3}{4}\displaystyle\int_0^1 xtu(t)dt$

6. $u(x) = e^{-x} + 2x + \dfrac{3}{2}\displaystyle\int_{-1}^0 xtu(t)dt$

7. $u(x) = 1 + x - \dfrac{1}{12}x^2 + \displaystyle\int_0^1 x^2tu(t)dt$

8. $u(x) = e^x - x^2 + \displaystyle\int_0^1 x^2tu(t)dt$

2.4   The Direct Computation Method

We next introduce an efficient traditional method for solving Fredholm integral equations of the second kind (1), called the direct computational method. Recall that our attention will be focused on separable or degenerate kernels $K(x,t)$ expressed in the form defined by (2). Without loss of generality, we may assume that the kernel of (1) can be expressed as

$$K(x,t) = g(x)h(t). \tag{102}$$

Accordingly, the equation (1) becomes

$$u(x) = f(x) + \lambda g(x)\int_a^b h(t)u(t)dt. \tag{103}$$

It is clear that the definite integral at the right hand side of (103) reveals that the integrand depends on one variable, namely the variable t. This means that the definite integral in the right side of (103) is equivalent to a numerical value α, where α is a constant. In other words, we may write

$$\alpha = \int_a^b h(t)u(t)dt. \tag{104}$$

It follows that equation (103) becomes

$$u(x) = f(x) + \lambda\alpha g(x). \tag{105}$$

It is thus obvious that the solution $u(x)$ is completely determined by (105) upon evaluating the constant α. This can be easily done by substituting

Eq. (105) into Eq. (104). We point out here that this approach is slightly different than other existing techniques in that we substitute (105) into (104) and not in (103) as used by other texts.

It is worth noting that the *direct computation method* determines the exact solution in a closed form, rather than a series form, provided that the constant α is evaluated. In addition, this method usually gives rise to a system of algebraic equations depending on the structure of the kernel, where sometimes we need to evaluate more than one constant as will be seen in Examples 3 and 4. For linear Fredholm integral equations, we obtain one value for α, or one value for each of α and β if these two constants are used. This is due to the fact that linear equation has a unique solution.

In what follows, we examine four illustrative examples by using the direct computation method.

Example 1. We will use the direct computation method to solve the following Fredholm integral equation

$$u(x) = \frac{5}{6}x + \frac{1}{2}\int_0^1 xt\,u(t)dt. \tag{106}$$

As indicated before we set

$$\alpha = \int_0^1 t\,u(t)dt, \tag{107}$$

where α is a constant that represents the numerical value of the integral (107). The equation (107) carries (106) into

$$u(x) = \left(\frac{5}{6} + \frac{1}{2}\alpha\right)x. \tag{108}$$

To determine α, we substitute (108) into (107) to obtain

$$\alpha = \int_0^1 \left(\frac{5}{6} + \frac{1}{2}\alpha\right)t^2 dt, \tag{109}$$

so that by integrating the right hand side and solving for α we find

$$\alpha = \frac{1}{3}. \tag{110}$$

Substituting (110) into (108) yields

$$u(x) = x, \tag{111}$$

the exact solution of the given Fredholm integral equation.

Example 2. We will use the direct computation method to solve the following Fredholm integral equation

$$u(x) = \sec^2 x - 1 + \int_0^{\frac{\pi}{4}} u(t)dt. \tag{112}$$

Proceeding as before we set

$$\alpha = \int_0^{\frac{\pi}{4}} u(t)dt, \tag{113}$$

and by substituting this into (112) yields

$$u(x) = \sec^2 x - 1 + \alpha. \tag{114}$$

Inserting (114) into (113) we find

$$\alpha = \int_0^{\frac{\pi}{4}} \left(\sec^2 t - 1 + \alpha \right) dt, \tag{115}$$

so that

$$\alpha = 1. \tag{116}$$

Substituting (116) into (114) gives

$$u(x) = \sec^2 x, \tag{117}$$

the exact solution of the Fredholm integral equation of Example 2.

Example 3. We will use the direct computation method to solve the following Fredholm integral equation

$$u(x) = -8x - 6x^2 + \int_0^1 \left(20xt^2 + 12x^2t \right) u(t)dt. \tag{118}$$

Noting that the kernel here is separable and consists of two terms, we can rewrite Eq. (118) as

$$u(x) = -8x - 6x^2 + 20x \int_0^1 t^2 u(t)dt + 12x^2 \int_0^1 tu(t)dt. \tag{119}$$

In a manner parallel to the preceding example, we set

$$\alpha = \int_0^1 t^2 u(t)dt, \tag{120}$$

and

$$\beta = \int_0^1 tu(t)dt, \tag{121}$$

where α and β are constants. Consequently, Eq. (119) can be expressed in the form

$$u(x) = (20\alpha - 8)\,x + (12\beta - 6)\,x^2. \tag{122}$$

Substituting (122) into (120) and (121) we obtain

$$\alpha = \int_0^1 \left[(20\alpha - 8)\,t + (12\beta - 6)\,t^2\right] t^2 dt, \tag{123}$$

and

$$\beta = \int_0^1 \left[(20\alpha - 8)\,t + (12\beta - 6)\,t^2\right] t\,dt. \tag{124}$$

Integrating the right hand side of equations (123) and (124) yields the system of equations

$$5\alpha + 3\beta \;=\; 4, \tag{125}$$

$$40\alpha + 12\beta \;=\; 25, \tag{126}$$

so that by solving this system we find

$$\alpha = \frac{9}{20}, \;\; \beta = \frac{7}{12}. \tag{127}$$

Inserting (127) into (122) gives

$$u(x) = x^2 + x. \tag{128}$$

Example 4. We will use the direct computation method to solve the following Fredholm integral equation

$$u(x) = 1 + 9x + 2x^2 + x^3 - \int_0^1 \left(20xt + 10x^2 t^2\right) u(t)dt. \tag{129}$$

Noting that the kernel here is separable and consists of two terms, we can rewrite Eq. (129) as

$$u(x) = 1 + 9x + 2x^2 + x^3 - 20x \int_0^1 tu(t)dt + 12x^2 \int_0^1 t^2 u(t)dt. \tag{130}$$

In a manner parallel to the preceding example, we set

$$\alpha = \int_0^1 tu(t)dt, \tag{131}$$

and

$$\beta = \int_0^1 t^2 u(t)dt, \tag{132}$$

where α and β are constants. Consequently, Eq. (130) can be expressed in the form

$$u(x) = 1 + (9 - 20\alpha)\,x + (2 - 10\beta)\,x^2 + x^3. \tag{133}$$

Substituting (133) into (131) and (132) we obtain

$$\alpha = \int_0^1 \left[1 + (9 - 20\alpha)\,t + (2 - 10\beta)\,t^2 + t^3\right] t\,dt, \tag{134}$$

and

$$\beta = \int_0^1 \left[1 + (9 - 20\alpha)\,t + (2 - 10\beta)\,t^2 + t^3\right] t^2\,dt. \tag{135}$$

Integrating the right hand side of equations (134) and (135) yields the system of equations

$$230\alpha + 75\beta \;=\; 126, \tag{136}$$

$$100\alpha + 60\beta \;=\; 63, \tag{137}$$

so that by solving this system we find

$$\alpha = \frac{9}{20}, \quad \beta = \frac{3}{10}. \tag{138}$$

Inserting (138) into (133) gives

$$u(x) = 1 - x^2 + x^3. \tag{139}$$

In closing this section, we point out that the direct computation method introduces a very direct technique to formally determine the solution of Fredholm integral equation. In this method, the Fredholm integral equation will be transformed into a more readily solvable integral. Moreover, the direct computation method introduces the exact solution in a closed form rather than a series form as in the case of the decomposition method or the variational iteration method. The other traditional methods, that will be discussed in the forthcoming sections, also determine the solution in a series form, but in a different approach than the decomposition method or the variational iteration method. We remark here that the direct computation method was introduced in this section in a slightly different manner than other texts.

Exercises 2.4

Solve the following Fredholm integral equations by using the *direct computation method*:

1. $u(x) = xe^x - x + \int_0^1 xu(t)dt.$

2. $u(x) = x^2 - \dfrac{25}{12}x + 1 + \int_0^1 xtu(t)dt.$

3. $u(x) = x \sin x - x + \int_0^{\pi/2} xu(t)dt.$

4. $u(x) = e^{2x} - \dfrac{1}{4}\left(e^2 + 1\right)x + \int_0^1 xtu(t)dt.$

5. $u(x) = \sec^2 x - \dfrac{\pi}{4} + \int_0^{\pi/4} u(t)dt.$

6. $u(x) = \sin(2x) - \dfrac{1}{2}x + \int_0^{\pi/4} xu(t)dt.$

7. $u(x) = x^2 - \dfrac{1}{3}x - \dfrac{1}{4} + \int_0^1 (x+2)u(t)dt.$

8. $u(x) = \sin x + \cos x - \dfrac{\pi}{2}x + \int_0^{\pi/2} xtu(t)dt.$

9. $u(x) = \sec x \tan x + x - \int_0^{\pi/3} xu(t)dt.$

10. $u(x) = x^2 - \dfrac{1}{6}x - \dfrac{1}{24} + \dfrac{1}{2}\int_0^1 (1+x-t)u(t)dt.$

11. $u(x) = \sin x - \dfrac{x}{4} + \dfrac{1}{4}\int_0^{\pi/2} xtu(t)dt.$

12. $u(x) = 1 + \int_{0+}^1 \ln(xt)u(t)dt, \quad 0 < x \le 1.$

13. $u(x) = \dfrac{9}{10}x^3 + \dfrac{1}{2}\int_0^1 x^3tu(t)dt.$

14. $u(x) = 1 + \dfrac{1}{2}\int_0^{\pi/4} \sec^2 x\, u(t)dt.$

2.5   The Successive Approximations Method

In this method, we replace the unknown function under the integral sign of the Fredholm integral equation of the second kind

$$u(x) = f(x) + \lambda \int_a^b K(x,t)u(t)dt, \ \ a \leq x \leq b, \tag{140}$$

by any *selective* real valued function $u_0(x)$, $a \leq x \leq b$. Accordingly, the first approximation $u_1(x)$ of the solution $u(x)$ is defined by

$$u_1(x) = f(x) + \lambda \int_a^b K(x,t)u_0(t)dt. \tag{141}$$

The second approximation of $u_2(x)$ of the solution $u(x)$ can be obtained by replacing $u_0(x)$ in (141) by the obtained approximation $u_1(x)$, hence we find

$$u_2(x) = f(x) + \lambda \int_a^b K(x,t)u_1(t)dt. \tag{142}$$

This process can be continued in the same manner to obtain the nth approximation. In other words, the various approximations of the solution $u(x)$ of (140) can be obtained in a recursive scheme given by

$$\begin{cases} u_0(x) & = \ \ \text{any selective real valued function} \\ \\ u_n(x) & = \ \ f(x) + \lambda \int_a^b K(x,t)u_{n-1}(t)dt, \ \ \ n \geq 1. \end{cases} \tag{143}$$

Even though we can select any real valued function for the zeroth approximation $u_0(x)$, the most commonly selected functions for $u_0(x)$ are 0, 1 or x. At the limit, the solution $u(x)$ is obtained by

$$u(x) = \lim_{n \to \infty} u_n(x), \tag{144}$$

so that the resulting solution $u(x)$ is independent of the choice of $u_0(x)$.

It is important to distinguish between the recursive schemes used in the Adomian decomposition method and in the successive approximations method. In the decomposition method, we apply the approach to determine several components of the solution $u(x)$ where, in this case

$$u(x) = \sum_{n=0}^{\infty} u_n(x), \tag{145}$$

so that the zeroth component $u_0(x)$ is defined by all terms that are out of the integral sign or part of these terms if the modified version is used. However, in the successive approximations method, we apply the above recursive scheme (143) to determine various approximations of the solution $u(x)$ itself, and not components of $u(x)$. Further, we should note here that the zeroth approximation $u_0(x)$ is not defined but rather given by a selective function, and as a result the solution $u(x)$ is given by the formula (144).

The successive approximations method will be illustrated by the following examples.

Example 1. Consider the Fredholm integral equation

$$u(x) = e^x + e^{-1} \int_0^1 u(t)\, dt. \tag{146}$$

As indicated above we can select any real value function for the zeroth component, hence we set

$$u_0(x) = 0. \tag{147}$$

Substituting (147) into the right hand side of (146) we find

$$u_1(x) = e^x + e^{-1} \int_0^1 u_0(t)\, dt, \tag{148}$$

and this gives the first approximation of $u(x)$ by

$$u_1(x) = e^x. \tag{149}$$

Inserting (149) into (148) to replace $u_0(x)$ we obtain

$$u_2(x) = e^x + e^{-1} \int_0^1 e^t\, dt, \tag{150}$$

where by integration we determine the second approximation of $u(x)$ by

$$u_2(x) = e^x + 1 - e^{-1}. \tag{151}$$

Continuing in the same manner we find the third approximation of $u(x)$ given by

$$u_3(x) = e^x + 1 - e^{-2}. \tag{152}$$

Proceeding as before, we obtain the nth component

$$u_n(x) = e^x + 1 - e^{-(n-1)}, \quad n \geq 1. \tag{153}$$

Using (144), the solution $u(x)$ of (146) is given by

$$
\begin{aligned}
u(x) &= \lim_{n \to \infty} u_n(x), \\
&= \lim_{n \to \infty} \left(e^x + 1 - e^{-(n-1)} \right) \\
&= e^x + 1,
\end{aligned}
\tag{154}
$$

obtained upon evaluating the limit as $n \to \infty$.

Example 2. We next consider the Fredholm integral equation

$$
u(x) = x + \lambda \int_0^1 xt u(t) dt.
\tag{155}
$$

The zeroth approximation may by selected by

$$
u_0(x) = 0,
\tag{156}
$$

where by substituting this in the right hand side of (155) the first approximation

$$
u_1(x) = x,
\tag{157}
$$

follows immediately. Proceeding in the same manner we find that

$$
u_2(x) = x + \lambda \int_0^1 xt^2 dt,
\tag{158}
$$

so that

$$
u_2(x) = x + \frac{\lambda}{3} x.
\tag{159}
$$

In a similar manner we obtain

$$
u_3(x) = x + \lambda \int_0^1 xt \left(1 + \frac{\lambda}{3} \right) tdt,
\tag{160}
$$

which yields

$$
u_3(x) = x + \frac{\lambda}{3} x + \frac{\lambda^2}{9} x.
\tag{161}
$$

Generally we obtain for the nth approximation

$$
u_n(x) = x + \frac{\lambda}{3} x + \frac{\lambda^2}{9} x + \cdots + \frac{\lambda^{n-1}}{3^{n-1}} x, \quad n \geq 1.
\tag{162}
$$

Consequently, the solution $u(x)$ of (155) is given by

$$
\begin{aligned}
u(x) &= \lim_{n\to\infty} u_n(x), \\
&= \lim_{n\to\infty}\left(x + \frac{\lambda}{3}x + \frac{\lambda^2}{9}x + \cdots\right) \qquad (163) \\
&= \frac{3}{3-\lambda}x, \quad 0 < \lambda < 3.
\end{aligned}
$$

To show that $u(x)$ obtained in (163) does not depend on the selection of $u_0(x)$, we will solve the equation (155) by selecting

$$
u_0(x) = x. \qquad (164)
$$

Using the new selection of $u_0(x)$ in the right hand side of (155) the first approximation

$$
u_1(x) = x + \frac{\lambda}{3}x, \qquad (165)
$$

is readily obtained. Proceeding as before we thus obtain

$$
u_2(x) = x + \lambda \int_0^1 xt\left(t + \frac{\lambda}{3}t\right) dt, \qquad (166)
$$

which gives

$$
u_2(x) = x + \frac{\lambda}{3}x + \frac{\lambda^2}{9}x. \qquad (167)
$$

In a parallel manner we find

$$
u_n(x) = x + \frac{\lambda}{3}x + \frac{\lambda^2}{3^2}x + \cdots + \frac{\lambda^n}{3^n}x, \quad n \geq 1. \qquad (168)
$$

Accordingly, we obtain

$$
u(x) = \frac{3}{3-\lambda}x, \quad 0 < \lambda < 3, \qquad (169)
$$

which is consistent with the same result obtained above in (163).

Exercises 2.5

Solve the following Fredholm integral equations by using the *successive approximations method*:

1. $u(x) = \dfrac{11}{12}x + \dfrac{1}{4}\displaystyle\int_0^1 xtu(t)dt.$

2. $u(x) = \dfrac{6}{7}x^3 + \dfrac{5}{7}\displaystyle\int_0^1 x^3tu(t)dt.$

3. $u(x) = \dfrac{13}{3}x - \dfrac{1}{4}\displaystyle\int_0^1 xtu(t)dt.$

4. $u(x) = 1 + \displaystyle\int_0^1 xu(t)dt.$

5. $u(x) = \sin x + \displaystyle\int_0^{\pi/2} \sin x \cos t\, u(t)dt.$

6. $u(x) = -\dfrac{1}{2} + \sec^2 x + \dfrac{1}{2}\displaystyle\int_0^{\pi/4} u(t)dt.$

7. $u(x) = -\dfrac{1}{4} + \sec x \tan x + \dfrac{1}{4}\displaystyle\int_0^{\pi/3} u(t)dt.$

8. $u(x) = \cosh x + \left(1 - e^{-1}\right)x - \displaystyle\int_0^1 xtu(t)dt.$

9. $u(x) = e^x - (\sinh 1)\,x + \dfrac{1}{2}\displaystyle\int_{-1}^1 xu(t)dt.$

10. $u(x) = \dfrac{1}{4}x + \sin x - \dfrac{1}{4}\displaystyle\int_0^{\pi/2} xu(t)dt.$

2.6 The Method of Successive Substitutions

This method introduces the solution of the integral equation in a series form through evaluating single integral and multiple integrals as well. The computational work needed in this method is huge compared with other techniques.

In this method, we set $x = t$ and $t = t_1$ in the Fredholm integral equation

$$u(x) = f(x) + \lambda \int_a^b K(x,t)u(t)dt, \quad a \le x \le b, \tag{170}$$

to obtain

$$u(t) = f(t) + \lambda \int_a^b K(t,t_1)u(t_1)dt_1. \tag{171}$$

Replacing $u(t)$ in the right hand side of (170) by its obtained value given by (171) yields

$$\begin{aligned} u(x) &= f(x) + \lambda \int_a^b K(x,t)f(t)dt \\ &+ \lambda^2 \int_a^b K(x,t) \int_a^b K(t,t_1)u(t_1)dt_1 dt. \end{aligned} \tag{172}$$

Substituting $x = t_1$ and $t = t_2$ in (170) we obtain

$$u(t_1) = f(t_1) + \lambda \int_a^b K(t_1, t_2)u(t_2)dt_2. \tag{173}$$

Substituting the value of $u(t_1)$ obtained in (173) into the right hand side of (172) leads to

$$u(x) = f(x) + \lambda \int_a^b K(x, t)f(t)dt$$
$$+ \lambda^2 \int_a^b \int_a^b K(x, t)K(t, t_1)f(t_1)dt_1 dt$$
$$+ \lambda^3 \int_a^b \int_a^b \int_a^b K(x, t)K(t, t_1)K(t_1, t_2)u(t_2)dt_2 dt_1 dt. \tag{174}$$

Accordingly, the general series form for $u(x)$ can be written as

$$u(x) = f(x) + \lambda \int_a^b K(x, t)f(t)dt$$
$$+ \lambda^2 \int_a^b \int_a^b K(x, t)K(t, t_1)f(t_1)dt_1 dt$$
$$+ \lambda^3 \int_a^b \int_a^b \int_a^b K(x, t)K(t, t_1)K(t_1, t_2)f(t_2)dt_2 dt_1 dt, \tag{175}$$

and so on. We note that the series solution given in (175) converges uniformly in the interval $[a, b]$ if $\lambda M(b - a) < 1$ where $|K(x, t)| \leq M$. The proof of the theorem appears in the texts [16], [19], [20] and others. We remark here that in this method the unknown function $u(x)$ is replaced by the given function $f(x)$ that makes the evaluation of the several multiple integrals possible and easily computable. This substitution of $u(x)$ occurs several times through the integrals and hence this is why it is called the method of successive substitutions. The technique will be illustrated by discussing the following examples.

Example 1. We solve the following Fredholm integral equation

$$u(x) = \frac{23}{6}x + \frac{1}{8}\int_0^1 xtu(t)dt, \tag{176}$$

by using the method of successive substitutions.
Substituting $\lambda = \frac{1}{8}$, $f(x) = \frac{23}{6}x$, and $K(x, t) = xt$ into (175) yields

$$u(x) = \frac{23}{6}x + \frac{1}{8}\int_0^1 \frac{23}{6}xt^2 dt + \frac{1}{8^2}\int_0^1 \int_0^1 \frac{23}{6}xt_1{}^2 t^2 dt_1 dt + \cdots, \tag{177}$$

or equivalently

$$u(x) = \frac{23}{6}x\left[1 + \frac{1}{24} + \frac{1}{576} + \cdots\right], \tag{178}$$

so that we obtain the solution

$$u(x) = 4x, \tag{179}$$

upon evaluating the sum of the geometric series.

Example 2. We next solve the Fredholm integral equation

$$u(x) = 1 + \frac{1}{4}\int_0^{\pi/2}\cos x\, u(t)dt, \tag{180}$$

by using the method of successive substitutions.
Substituting $\lambda = \frac{1}{4}$, $f(x) = 1$, and $K(x,t) = \cos x$ into (175) yields

$$u(x) = 1 + \frac{1}{4}\int_0^{\pi/2}\cos x\, dt + \frac{1}{16}\int_0^{\pi/2}\int_0^{\pi/2}\cos x\,\cos t\, dt_1\, dt + \cdots, \tag{181}$$

and this will yield

$$u(x) = 1 + \left(\frac{\pi}{8}\cos x + \frac{\pi}{32}\cos x + \cdots\right), \tag{182}$$

which gives the exact solution

$$u(x) = 1 + \frac{\pi}{6}\cos x, \tag{183}$$

obtained upon using the sum of the infinite geometric series.

Exercises 2.6

Solve the following Fredholm integral equations by using the *successive substitutions method*:

1. $u(x) = \frac{11}{6}x + \frac{1}{4}\int_0^1 xtu(t)dt.$

2. $u(x) = 1 - \frac{1}{4}\int_0^{\pi/2}\cos xu(t)dt.$

3. $u(x) = \frac{7}{12}x + 1 + \frac{1}{2}\int_0^1 xtu(t)dt.$

4. $u(x) = \cos x + \frac{1}{2}\int_0^{\pi/2}\sin xu(t)dt.$

5. $u(x) = \dfrac{7}{8}x^2 + \dfrac{1}{2}\displaystyle\int_0^1 x^2 t u(t)dt.$

6. $u(x) = \dfrac{9}{10}x^3 + \dfrac{1}{2}\displaystyle\int_0^1 x^3 t u(t)dt.$

7. $u(x) = \sin x + \dfrac{1}{2}\displaystyle\int_0^{\pi/2} \cos x u(t)dt.$

8. $u(x) = 1 + \dfrac{1}{2}\displaystyle\int_0^{\pi/2} \sin x u(t)dt.$

9. $u(x) = 1 + \dfrac{1}{2}\displaystyle\int_0^{\pi/4} \sec^2 x\, u(t)dt.$

10. $u(x) = 1 + \dfrac{1}{5}\displaystyle\int_0^{\pi/3} \sec x \tan x\, u(t)dt.$

2.7 Comparison between Alternative Methods

Having finished the mathematical analysis of the methods that handle Fredholm integral equations, we are now ready to carry out a comparison between these methods. When it comes to selecting a preferable method among the five methods for solving linear Fredholm integral equations, we cannot recommend a specific method. However, we found that if the separable kernel $K(x, t)$ of the integral equation consists of a polynomial of one or two terms only, the *direct computation method* might be the best choice because it provides the exact solution with the minimum volume of calculations. For other types of kernels, and if in addition the nonhomogeneous part $f(x)$ is a polynomial of more than two terms we found that the *Adomian decomposition method* or the *Variational Iteration Method*, are proved to be effective, reliable and produces a rapid convergent series for the solution. The series obtained by using the decomposition method may give the solution in a closed form or we may obtain an approximation of high accuracy level by using a truncated series for concrete problems.

It is worth noting that the decomposition method expands the solution $u(x)$ about a function, instead of a point as in Taylor theorem.

To compare the *decomposition method* with the *successive approximation method*, it is clear that the decomposition method is easier in that we always integrate few terms to obtain the successive components, whereas in the other method we integrate many terms to evaluate the successive approximations after selecting the zeroth approximation. The two methods give the solution in a series form.

In addition, we point out that the *method of successive substitutions* suffers from the huge size of calculations in evaluating the several multiple integrals especially if the function $f(x)$ is a trigonometric, logarithmic or exponential function. However, the method is directly based on substituting the unknown function $u(x)$ under the integral sign by the given function $f(x)$.

It is to be noted that, for a first course in integral equations, we introduced five methods only to handle Fredholm integral equations, noting that other traditional techniques are left for a further study.

To achieve our goal of the comparison between these methods, we demonstrate this comparison by discussing the following example by using all various methods.

Example 1. We solve the following example

$$u(x) = \frac{5}{6}x + \frac{1}{2}\int_0^1 xtu(t)dt, \tag{184}$$

by using the five alternative methods discussed before.

(a) The Adomian Decomposition Method: In this method, we have to set the zeroth component u_0 by all terms outside the integral sign, hence we have

$$u_0(x) = \frac{5}{6}x. \tag{185}$$

Using (185) we obtain the first component $u_1(x)$ by

$$u_1(x) = \frac{1}{2}x\int_0^1 \frac{5}{6}t^2dt, \tag{186}$$

so that

$$u_1(x) = \frac{5}{6^2}x. \tag{187}$$

Proceeding in the same manner we can easily obtain

$$u_2(x) = \frac{5}{6^3}x. \tag{188}$$

Noting that in the decomposition method we have

$$u(x) = u_0 + u_1 + u_2 + u_3 + \cdots, \tag{189}$$

so that

$$u(x) = \frac{5}{6}x\left(1 + \frac{1}{6} + \frac{1}{6^2} + \frac{1}{6^3} + \cdots\right), \tag{190}$$

and this gives the exact solution

$$u(x) = x, \tag{191}$$

obtained by evaluating the sum of the infinite geometric series.

(b) The Variational Iteration Method: Differentiate both sides gives

$$u'(x) = \frac{5}{6} + \frac{1}{2}\int_0^1 tu(t)\, dt, u(0) = 0. \tag{192}$$

The correction functional for this equation is given by

$$u_{n+1}(x) = u_n(x) - \int_0^x \left(u_n'(\xi) - \frac{5}{6} - \int_0^1 ru_n(r)\, dr \right)\, d\xi. \tag{193}$$

The initial condition $u(0) = 0$ is obtained by substituting $x = 0$ into (96). Using this selection into the correction functional gives the following successive approximations

$$
\begin{aligned}
u_0(x) &= 0, \\
u_1(x) &= \frac{5}{6}x, \\
u_2(x) &= \frac{35}{36}x, \\
u_3(x) &= \frac{215}{216}x, \\
u_4(x) &= \frac{1295}{1296}x, \\
&= \vdots, \\
u_n(x) &= \frac{6^n-1}{6^n}x.
\end{aligned}
\tag{194}
$$

This converges to the exact solution

$$u(x) = x. \tag{195}$$

(c) The Direct Computation Method: As discussed earlier we set

$$\alpha = \int_0^1 t\, u(t)\, dt, \tag{196}$$

where α is a constant that represents the numerical value of the integral (196). The equation (196) carries (184) into

$$u(x) = \left(\frac{5}{6} + \frac{1}{2}\alpha \right) x. \tag{197}$$

To determine α, we substitute (197) into (196) to obtain

$$\alpha = \int_0^1 \left(\frac{5}{6} + \frac{1}{2}\alpha \right) t^2 dt, \tag{198}$$

so that by integrating the right hand side and solving for α we find

$$\alpha = \frac{1}{3}. \tag{199}$$

Substituting (199) into (197) yields

$$u(x) = x, \tag{200}$$

the exact solution of the equation (184).

(d) The Successive Approximations Method: In this method we select the zeroth approximation by

$$u_0(x) = 0. \tag{201}$$

Following the technique that was discussed above, the other approximations of the solution $u(x)$ can be easily obtained by

$$u_1(x) = \frac{5}{6}x, \tag{202}$$

$$u_2(x) = \left(\frac{5}{6} + \frac{5}{6^2}\right)x, \tag{203}$$

$$u_3(x) = \left(\frac{5}{6} + \frac{5}{6^2} + \frac{5}{6^3}\right)x, \tag{204}$$

and so on. Accordingly, the nth component is given by

$$u_n(x) = \left(\frac{5}{6} + \frac{5}{6^2} + \frac{5}{6^3} + \frac{5}{6^4} + \cdots + \frac{5}{6^n}\right)x, \quad n \geq 1. \tag{205}$$

Consequently we find

$$
\begin{aligned}
u(x) &= \lim_{n\to\infty} u_n(x), \\
&= \lim_{n\to\infty} \frac{5}{6}x\left(1 + \frac{1}{6} + \frac{1}{6^2} + \frac{1}{6^3} + \frac{1}{6^4} + \cdots\right) \\
&= x,
\end{aligned} \tag{206}
$$

the same result obtained above.

(e) The Method of Successive Substitutions: In this method we have to set $K(x,t) = xt$, $\lambda = \frac{1}{2}$ and $f(x) = \frac{5}{6}x$, hence we have

$$
\begin{aligned}
u(x) &= \frac{5}{6}x + \frac{1}{2}\int_0^1 \frac{5}{6}xt^2 dt + \frac{1}{4}\int_0^1 \int_0^1 \frac{5}{6}x^2 tt_1^2 dt_1 dt + \cdots \\
&= \frac{5}{6}x\left(1 + \frac{1}{6} + \frac{1}{6^2} + \cdots\right) \\
&= x.
\end{aligned}
\tag{207}
$$

This confirms our belief that the Adomian decomposition method, the variational iteration method, and the direct computation method reduce the size of calculations and provide improvements if compared with the other traditional techniques. In addition, the Adomian decomposition method gives the components of the solution $u(x)$. However, the variational iteration method gives the successive approximations of the solution $u(x)$ instead of the components.

2.8 Homogeneous Fredholm Integral Equations

In this section we will study the homogeneous Fredholm integral equation with separable kernel given by

$$
u(x) = \lambda \int_a^b K(x,t)u(t)\, dt,
\tag{208}
$$

obtained from (1) by setting $f(x) = 0$. It is easily seen that the trivial solution $u(x) = 0$ is a solution of the homogeneous Fredholm integral equation (208). In this study our goal will be focused on finding nontrivial solutions to (208) if exist. We can achieve our goal by introducing the technique that will enable us to determine the nontrivial solutions to (208). Generally speaking, the homogeneous Fredholm integral equation with separable kernel may have nontrivial solutions. Our approach in obtaining these desired solutions will be based mainly on the *direct computation method* that was employed effectively for nonhomogeneous Fredholm integral equations. We point out that Adomian decomposition method is not applicable for the homogeneous Fredholm integral equations. This may be related to the fact that the nonhomogeneous part $f(x)$ does not exist in this type of problems, and therefore the zeroth component $u_0(x)$ cannot be defined.

We recall that the direct computation method reduces the equation to an algebraic equation if the kernel consists of one term only, or to a system of algebraic equations if the kernel contains many separable terms. Additional discussions will be required for determining possible values of λ that will give rise to the nontrivial solutions as will be discussed soon.

Without loss of generality, we may assume a one term kernel given by

$$K(x,t) = g(x)\,h(t), \tag{209}$$

so that (208) becomes

$$u(x) = \lambda g(x) \int_a^b h(t)u(t)\,dt. \tag{210}$$

Using the direct computation method we set

$$\alpha = \int_a^b h(t)u(t)\,dt, \tag{211}$$

so that (210) becomes

$$u(x) = \lambda \alpha g(x). \tag{212}$$

We note that $\alpha = 0$ gives the trivial solution $u(x) = 0$, by using Eq. (212), which is not our desired goal in this study. However, to determine the nontrivial solutions of (208), we need to determine the values of the parameter λ by considering $\alpha \neq 0$. This can be done by substituting (212) into (211) to obtain

$$\alpha = \lambda \alpha \int_a^b h(t)g(t)dt, \tag{213}$$

or equivalently

$$1 = \lambda \int_a^b h(t)g(t)dt \tag{214}$$

which gives a numerical value for $\lambda \neq 0$ by evaluating the definite integral in (214). Having evaluated λ, the nontrivial solution given by (212) is determined.

For separable kernels that contain more than one term, the method reduces the homogeneous Fredholm integral equation to a system of algebraic equations as will be seen from the examples below.

In closing this section we point out that the particular nonzero values of λ that result from solving the algebraic system of equations are called the *eigenvalues* of the kernel. Moreover, substituting the obtained values of λ in (212) gives the usually called *eigenfunctions* of the equation which are the nontrivial solutions of (208).

The following illustrative examples will be used to explain the technique introduced above and the concept of *eigenvalues* and *eigenfunctions*.

Example 1. We first solve the homogeneous Fredholm integral equation with one term kernel

$$u(x) = \lambda \int_0^1 x\, u(t)dt. \tag{215}$$

As indicated above, (215) becomes

$$u(x) = \lambda \alpha x, \tag{216}$$

where

$$\alpha = \int_0^1 u(t)dt. \tag{217}$$

Substituting (216) into (217) yields

$$\alpha = \lambda \alpha \int_0^1 t\, dt, \tag{218}$$

which gives

$$\alpha = \frac{1}{2}\lambda \alpha, \tag{219}$$

so that the eigenvalue of the kernel

$$\lambda = 2, \tag{220}$$

obtained by noting that $\alpha \neq 0$. Substituting (220) into (219) yields

$$\alpha = \alpha, \tag{221}$$

which indicates that α is an arbitrary constant. Using (220) and (221) in (216) leads to the eigenfunction of the equation given by

$$u(x) = 2\alpha x, \tag{222}$$

obtained upon using (216).

Example 2. We next solve the homogeneous Fredholm integral equation

$$u(x) = \frac{2}{\pi}\lambda \int_0^\pi \cos(x+t)u(t)dt. \tag{223}$$

The equation (223) can be rewritten as

$$u(x) = \frac{2}{\pi}\lambda \cos x \int_0^\pi \cos t\, u(t)dt - \frac{2}{\pi}\lambda \sin x \int_0^\pi \sin t\, u(t)dt, \tag{224}$$

or by

$$u(x) = \frac{2}{\pi}\lambda \left(\alpha \cos x - \beta \sin x\right), \tag{225}$$

where

$$\alpha = \int_0^\pi \cos t\, u(t)\, dt, \tag{226}$$

$$\beta = \int_0^\pi \sin t\, u(t)\, dt. \tag{227}$$

Substituting (225) into (226) and (227) and integrating yield

$$\alpha = \lambda\alpha, \tag{228}$$

and

$$\beta = -\lambda\beta. \tag{229}$$

For $\alpha \neq 0$ and $\beta \neq 0$, we obtain the eigenvalues

$$\lambda_1 = 1, \quad \text{and} \quad \lambda_2 = -1. \tag{230}$$

Substituting $\lambda_1 = 1$ in (228) and (229) yields

$$\alpha = \alpha, \quad \text{and} \quad \beta = 0, \tag{231}$$

which gives the eigenfunction corresponding to $\lambda_1 = 1$ by

$$u_1(x) = \frac{2}{\pi}\alpha \cos x, \tag{232}$$

obtained upon using (225), where α is an arbitrary constant.
Similarly, substituting $\lambda_2 = -1$ in (228) and (229) yields

$$\alpha = 0, \quad \text{and} \quad \beta = \beta, \tag{233}$$

which gives the second eigenfunction corresponding to $\lambda_2 = -1$ by

$$u_2(x) = \frac{2}{\pi}\beta \sin x, \tag{234}$$

obtained upon using (225), where β is an arbitrary constant.

Example 3. We finally solve the homogeneous Fredholm integral equation with two-term kernel

$$u(x) = \lambda \int_0^1 (6x - 2t)u(t)dt. \tag{235}$$

Equation (235) can be rewritten as

$$u(x) = 6\lambda\alpha x - \beta\lambda, \tag{236}$$

where

$$\alpha = \int_0^1 u(t)\, dt, \tag{237}$$

$$\beta = \int_0^1 2tu(t)\, dt. \tag{238}$$

Substituting (236) into (237) and (238) and integrating yield

$$(1 - 3\lambda)\alpha + \lambda\beta = 0, \tag{239}$$

and

$$-4\lambda\alpha + (1 + \lambda)\beta = 0. \tag{240}$$

For $\alpha \neq 0$ and $\beta \neq 0$, we obtain the eigenvalues

$$\lambda_1 = \lambda_2 = 1. \tag{241}$$

Substituting $\lambda_1 = 1$ in (239) and (240) yields

$$\beta = 2\alpha. \tag{242}$$

Consequently the eigenfunctions corresponding to $\lambda_1 = \lambda_2 = 1$ are given by

$$u_1(x) = u_2(x) = 6\alpha x - 2\alpha, \tag{243}$$

obtained upon using (236).

Exercises 2.8

Find the nontrivial solutions for following homogeneous Fredholm integral equations by using the *eigenvalues* and *eigenfunctions* concepts

1. $u(x) = \lambda \int_0^1 2tu(t)dt.$

2. $u(x) = \lambda \int_0^1 4xu(t)dt.$

3. $u(x) = \lambda \int_0^1 xe^t u(t)dt.$

4. $u(x) = \lambda \int_0^{\pi/2} \cos x \sin t\, u(t)dt.$

5. $u(x) = \dfrac{2}{\pi}\lambda \int_0^{\pi} \sin(x + t)u(t)dt.$

6. $u(x) = \dfrac{2}{\pi}\lambda \int_0^{\pi} \cos(x - t)u(t)dt.$

7. $u(x) = \lambda \displaystyle\int_0^{\pi/3} \sec x \tan t \, u(t) dt.$

8. $u(x) = \lambda \displaystyle\int_0^{\pi/4} \sec^2 x \, u(t) dt.$

9. $u(x) = \lambda \displaystyle\int_0^1 \sin^{-1} x u(t) dt.$

10. $u(x) = \lambda \displaystyle\int_0^1 (3 - \frac{3}{2}x) t u(t) dt.$

2.9 Fredholm Integral Equations of the First Kind

It was stated before that in Fredholm integral equations of the first kind, the unknown function $u(x)$ appears only inside the integral sign. The standard form of the Fredholm integral equations of the first kind reads

$$f(x) = \lambda \int_a^b K(x,t) u(t) \, dt, x \in D, \tag{244}$$

where $f(x)$ is the data. The occurrence of $u(x)$ under the integral sign makes it difficult to apply the aforementioned method. It was proved in the literature that the data function $f(x)$ must lie in the range of the kernel $K(x,t)$. This means that if we set the kernel by $K(x,t) = e^x \cos t$, then for any $u(x)$, the resulting data $f(x)$ must be a multiple of e^x, otherwise the solution $u(x)$ does not exist. The Fredholm integral equations of the first kind appear in many physical models such as radiography, spectroscopy, cosmic radiation, image processing and in the theory of signal processing.

In the literature, the Fredholm integral equations of the first kind is considered ill-posed problem, and this indicates that this first kind Fredholm equation may have no solution, or if a solution exists it is not unique and may not depend continuously on the data [40].

There are many methods that were used to investigate the Fredholm integral equations of the first kind, analytically and numerically. However, in this text we will use the *method of regularization* that transforms first kind equation to second kind equation. Having converted the equation from the first kind to the second kind enables us to apply any appropriate method that were presented in this chapter.

2.9.1 The Method of Regularization

The method of regularization was introduced in [33, 40]. The method of regularization converts the linear Fredholm integral equation of the first kind

$$f(x) = \int_a^b K(x,t)u(t)\, dt, x \in D, \tag{245}$$

to the Fredholm integral equation of the second kind in the form

$$\epsilon u_\epsilon(x) = f(x) - \int_a^b K(x,t)u_\epsilon(t)\, dt, \tag{246}$$

or equivalently

$$u_\epsilon(x) = \frac{1}{\epsilon}f(x) - \frac{1}{\epsilon}\int_a^b K(x,t)u_\epsilon(t)\, dt, \tag{247}$$

where ϵ is a small positive parameter. It is obvious that the solution u_ϵ of equation (247) converges to the solution $u(x)$ of (245) as $\epsilon \to 0$.

In what follows we will present two illustrative examples. Our focus will be on transforming the first kind equation to a second kind equation by using the method of regularization, and hence we can use any appropriate method.

Example 1. Combine the method of regularization and the direct computation method to solve the Fredholm integral equation of the first kind

$$\frac{1}{2}e^{2x} = \int_0^{\frac{1}{2}} e^{2x-2t}\, u(t)\, dt. \tag{248}$$

Using the method of regularization, Eq. (248) becomes

$$u_\epsilon(x) = \frac{1}{2\epsilon}e^{2x} - \frac{1}{\epsilon}\int_0^{\frac{1}{2}} e^{2x-2t}u_\epsilon(t)\, dt. \tag{249}$$

Using the direct computation method, Eq. (249) becomes

$$u_\epsilon(x) = (\frac{1}{2\epsilon} - \frac{\alpha}{\epsilon})e^{2x}, \tag{250}$$

where

$$\alpha = \int_0^{\frac{1}{2}} e^{-2t}u_\epsilon(t)\, dt. \tag{251}$$

To determine α, we substitute (250) into (251), integrate and solving to find that

$$\alpha = \frac{1}{2+4\epsilon}. \tag{252}$$

This in turn gives

$$u_\epsilon(x) = \frac{e^{2x}}{1 + 2\epsilon}. \tag{253}$$

The exact solution is given by

$$u(x) = \lim_{\epsilon \to 0} u_\epsilon(x) = e^{2x}. \tag{254}$$

Example 2.   Combine the method of regularization and the Adomian decomposition method to solve the Fredholm integral equation of the first kind

$$\frac{1}{5}x = \int_0^1 xt\, u(t)\, dt. \tag{255}$$

Using the method of regularization, Eq. (255) can be transformed to

$$u_\epsilon(x) = \frac{1}{5\epsilon}x - \frac{1}{\epsilon}\int_0^1 xt\, u_\epsilon(t)\, dt. \tag{256}$$

To use the Adomian decomposition method, we first select $u_{\epsilon_0}(x) = 0$. Consequently, we obtain the following approximations

$$
\begin{aligned}
u_{\epsilon_0}(x) &= \tfrac{1}{5\epsilon}x, \\
u_{\epsilon_1}(x) &= -\tfrac{1}{15\epsilon^2}x, \\
u_{\epsilon_2}(x) &= \tfrac{1}{45\epsilon^3}x, \\
u_{\epsilon_3}(x) &= -\tfrac{1}{135\epsilon^4}x, \\
&\vdots
\end{aligned}
\tag{257}
$$

and so on. Based on this we obtain the solution

$$u_\epsilon(x) = \frac{3}{5(1 + 3\epsilon)}x. \tag{258}$$

The exact solution $u(x)$ of (255) can be obtained by

$$u(x) = \lim_{\epsilon \to 0} u_\epsilon(x) = \frac{3}{5}x. \tag{259}$$

It is interesting to point out that another solution to this equation is given by

$$u(x) = x^3. \tag{260}$$

This is normal to get more than one solution because the Fredholm integral equation of the first kind is ill-posed problem.

Exercises 2.9

Combine the regularization method with any other method to solve the Fredholm integral equations of the first kind

1. $\dfrac{1}{3}x = \displaystyle\int_0^1 xtu(t)dt$

2. $\dfrac{1}{2}e^{-x} = \displaystyle\int_0^{\frac{1}{2}} e^{t-x}u(t)dt$

3. $\dfrac{3}{4}x = \displaystyle\int_0^1 xt^2u(t)dt$

4. $\dfrac{2}{5}x^2 = \displaystyle\int_{-1}^1 x^2t^2u(t)dt$

5. $\dfrac{\pi}{2}\cos x = \displaystyle\int_0^\pi \cos(x-t)u(t)dt$

6. $-\dfrac{\pi}{2}\cos x = \displaystyle\int_0^\pi \sin(x-t)u(t)dt$

Chapter 3

Volterra Integral Equations

3.1 Introduction

In this chapter we will be concerned with the nonhomogeneous Volterra integral equation of the second kind of the form

$$u(x) = f(x) + \lambda \int_0^x K(x,t)u(t)dt, \tag{1}$$

where $K(x,t)$ is the kernel of the integral equation, and λ is a parameter. As indicated earlier the limits of integration for the Volterra integral equations are functions of x and not constants as in Fredholm integral equations. The kernel in equation (1) will be considered a separable kernel as discussed before in the previous chapter. Our concern will be on applying various methods to determine the solution $u(x)$ of (1) and not on the abstract theorems related to the existence, uniqueness of the solution or the convergence concept. These important concepts can be found in a variety of integral equations texts.

We discussed in Section 1.5 the technique that converts initial value problems to Volterra integral equations. In the following we will discuss several methods that handle successfully the linear Volterra integral equations in a manner parallel to our approach in discussing Chapter 2. There is a variety on analytic and numerical techniques, traditional and new, that are usually used in studying Volterra integral equations. Accordingly we will first start with the recent methods.

3.2   The Adomian Decomposition Method

As stated earlier, Adomian developed the *Adomian decomposition method* or simply the *decomposition method* that proved to work for all types of differential, integral and integro-differential equations, linear or nonlinear. The method was introduced by Adomian in his books [1] and [2] and other related research papers such as [3] and [4]. The focus of the two books was mainly on ordinary and partial differential equations. We have seen from Chapter 2 that the decomposition method mostly establishes the solution in the form of a power series. The approach we will follow here is identical to the same approach that was implemented earlier in Chapter 2. In this method, the solution $u(x)$ will be decomposed into an infinite series of components, that will be determined, given by the series form

$$u(x) = \sum_{n=0}^{\infty} u_n(x), \tag{2}$$

with u_0 identified by all terms out of the integral sign, i.e.

$$u_0(x) = f(x). \tag{3}$$

Substituting (2) into (1) yields

$$\sum_{n=0}^{\infty} u_n(x) = f(x) + \lambda \int_0^x K(x,t) \left(\sum_{n=0}^{\infty} u_n(t) \right) dt, \tag{4}$$

which by using few terms of the expansion gives

$$
\begin{aligned}
u_0(x) + u_1(x) + u_2(x) + \cdots \; = f(x) \; &+ \lambda \int_0^x K(x,t) u_0(t) dt \\
&+ \lambda \int_0^x K(x,t) u_1(t) dt \\
&+ \lambda \int_0^x K(x,t) u_2(t) dt \\
&+ \lambda \int_0^x K(x,t) u_3(t) dt \\
&+ \cdots .
\end{aligned} \tag{5}
$$

The components $u_i(x), i \geq 0$ of the unknown function $u(x)$ are completely determined by using the recurrence manner

$$u_0(x) = f(x), \tag{6}$$

$$u_1(x) = \lambda \int_0^x K(x,t) u_0(t) dt, \tag{7}$$

$$u_2(x) = \lambda \int_0^x K(x,t)u_1(t)dt, \tag{8}$$

$$u_3(x) = \lambda \int_0^x K(x,t)u_2(t)dt, \tag{9}$$

and so on. The above discussed scheme for the determination of the components $u_i(x), i \geq 0$ of the solution $u(x)$ of Eq. (1) can be written in a recurrence relation by

$$u_0(x) = f(x), \tag{10}$$

$$u_{n+1}(x) = \lambda \int_0^x K(x,t)u_n(t)dt, \quad n \geq 0. \tag{11}$$

In view of (10) and (11), the components $u_i(x), i \geq 0$, follow immediately upon integrating the easily computable integrals. With these components determined, the solution $u(x)$ of (1) is readily determined in a series form upon using (2). As discussed before, the series obtained for $u(x)$ frequently provides the exact solution in a closed form if an exact solution exists as will be illustrated later. However, for concrete problems, where (2) cannot be evaluated, a truncated series $\sum_{n=0}^k u_n(x)$ is usually used to approximate the solution $u(x)$. It is to be noted here that, for numerical purposes, few terms of the obtained series usually provide the higher accuracy level of the approximation of the solution if compared with the existing numerical techniques.

It is interesting to recall that the decomposition method provides the solution of any style of equations in the form of a power series with easily computable components. In addition, applications have shown a very fast convergence of the series solution. The convergence concept of the decomposition technique was addressed extensively in [7] and by others, but it will not be discussed in this text.

We indicated earlier, that the decomposition technique proved to be powerful and reliable even if applied to nonlinear Volterra integral equations as will be discussed in the forthcoming Chapters.

The following illustrative examples will be discussed to explain the above outlined decomposition method.

Example 1. We first consider the Volterra integral equation

$$u(x) = 1 + \int_0^x u(t)\, dt. \tag{12}$$

It is clear that $f(x) = 1, \lambda = 1, K(x,t) = 1$. Using the decomposition series solution (2) and the recursive scheme (10) and (11) to determine the components u_n, $n \geq 0$, we find

$$u_0(x) = 1, \tag{13}$$

$$u_1(x) = \int_0^x u_0(t)dt$$

$$= \int_0^x dt = x, \tag{14}$$

$$u_2(x) = \int_0^x u_1(t)dt$$

$$= \int_0^x tdt = \frac{1}{2!}x^2, \tag{15}$$

and so on. Noting that

$$u(x) = u_0(x) + u_1(x) + u_2(x) + \cdots, \tag{16}$$

we can easily obtain the solution in a series form given by

$$u(x) = 1 + x + \frac{1}{2!}x^2 + \cdots, \tag{17}$$

and this converges to the closed form solution

$$u(x) = e^x, \tag{18}$$

obtained upon using the Taylor expansion for e^x.

Example 2. We next consider the Volterra integral equation

$$u(x) = x + \int_0^x (t - x)u(t)\, dt. \tag{19}$$

Proceeding as in Example 1, we set

$$u_0(x) = x, \tag{20}$$

$$u_1(x) = \int_0^x (t - x)u_0(t)dt$$

$$= \int_0^x t\,(t - x)\, dt = \frac{1}{3!}x^3, \tag{21}$$

$$u_2(x) = \int_0^x (t - x)u_1(t)dt$$

$$= \int_0^x -\frac{1}{3!}t^3\,(t - x)\, dt = \frac{1}{5!}x^5. \tag{22}$$

Consequently, the solution of (19) in a series form is given by

$$u(x) = x - \frac{1}{3!}x^3 + \frac{1}{5!}x^5 + \cdots \tag{23}$$

and in a closed form by

$$u(x) = \sin x, \tag{24}$$

obtained by using the Taylor expansion of $\sin x$.

Example 3. We consider here the Volterra integral equation

$$u(x) = 6x - x^3 + \frac{1}{2}\int_0^x t\, u(t)\, dt. \tag{25}$$

Applying the decomposition technique as discussed before we find

$$u_0(x) = 6x - x^3, \tag{26}$$

$$u_1(x) = \frac{1}{2}\int_0^x t u_0(t) dt$$

$$= \frac{1}{2}\int_0^x t\left(6t - t^3\right) dt = x^3 - \frac{1}{10}x^5, \tag{27}$$

$$u_2(x) = \frac{1}{2}\int_0^x t u_1(t) dt$$

$$= \frac{1}{2}\int_0^x t\left(t^3 - \frac{1}{10}t^5\right) dt = \frac{1}{10}x^5 - \frac{1}{140}x^7. \tag{28}$$

Consequently, the solution of (25) in a series form is given by

$$u(x) = (6x - x^3) + \left(x^3 - \frac{1}{10}x^5\right) + \left(\frac{1}{10}x^5 - \frac{1}{140}x^7\right) + \cdots, \tag{29}$$

where we can easily obtain the solution in a closed form given by

$$u(x) = 6x, \tag{30}$$

by eliminating the self-cancelling noise terms between various components of the solution $u(x)$.

It was discussed in Section 1.3 that it is not always possible to determine a solution in a closed form, but instead the solution obtained may be expressed in a series form. The difference between the closed form solution and the power series solution has been illustrated before in Section

1.3. However the series solution is usually employed for numerical approximations, and the more terms we obtain provide more accuracy in the approximation of the solution. In the next example we discuss again the solution expressed in a series form.

Example 4. Consider the Volterra integral equation

$$u(x) = 1 + \frac{1}{2} \int_0^x xt^2 u(t)\, dt. \tag{31}$$

Following the procedure used above, we find

$$u_0(x) \quad = \quad 1, \tag{32}$$

$$u_1(x) \quad = \quad \frac{1}{2} x \int_0^x t^2 dt$$

$$= \quad \frac{1}{6} x^4, \tag{33}$$

$$u_2(x) \quad = \quad \frac{1}{12} x \int_0^x t^6 dt$$

$$= \quad \frac{1}{84} x^8, \tag{34}$$

$$u_3(x) \quad = \quad \frac{1}{168} x \int_0^x t^{10} dt$$

$$= \quad \frac{1}{1848} x^{12}. \tag{35}$$

Consequently, the solution of (31) in a series form is given by

$$u(x) = 1 + \frac{1}{6} x^4 + \frac{1}{84} x^8 + \frac{1}{1848} x^{12} + \cdots, \tag{36}$$

so that a closed form for $u(x)$ does not appear obtainable.

Even though the decomposition method proved to be powerful and reliable, but it can be used sometimes in a more effective manner which we called the *modified decomposition method*. In a manner parallel to that used in Chapter 2, the volume of calculations will be reduced by evaluating only the first two components $u_0(x)$ and $u_1(x)$. The modified technique works for specific problems where the function $f(x)$ in (1) consists of at least two terms.

3.2.1 The Modified Decomposition Method

It is important to note that the modified decomposition method, that was introduced before in Chapter 2 for the Fredholm integral equations, is also applicable here. In Volterra integral equations where the nonhomogeneous part $f(x)$ in (1) consists of a polynomial that includes many terms, or in the case $f(x)$ contains a combination of polynomial and other trigonometric or transcendental functions, the modified decomposition method showed to work extremely well. As indicated earlier, the technique may minimize the volume of calculations needed when applying the standard decomposition method. To achieve our goal, we decompose the function $f(x)$ into two parts such as

$$f(x) = f_0(x) + f_1(x), \tag{37}$$

where $f_0(x)$ consists of one term only, or if needed of more terms in fewer other cases, and $f_1(x)$ includes the remaining terms of $f(x)$. Accordingly, Eq. (1) becomes

$$u(x) = f_0(x) + f_1(x) + \lambda \int_0^x K(x,t)u(t)dt. \tag{38}$$

Substituting the decomposition given by (2) into (38) and using few terms of the expansions we obtain

$$
\begin{aligned}
u_0(x) + u_1(x) + u_2(x) + \cdots \; = f_0(x) + f_1(x) \; &+ \lambda \int_0^x K(x,t)u_0(t)dt \\
&+ \lambda \int_0^x K(x,t)\,u_1(t)dt \\
&+ \lambda \int_0^x K(x,t)\,u_2(t)dt \\
&+ \lambda \int_0^x K(x,t)\,u_3(t)dt.
\end{aligned}
\tag{39}
$$

Consequently, the components $u_i(x), i \geq 0$ of the unknown function $u(x)$ can be completely determined in a modified recurrence relation if we assign $f_0(x)$ only to the component $u_0(x)$, whereas the component $f_1(x)$ will be added to the formula of the component $u_1(x)$ given before in Eq. (7). In other words, the modified recurrence relation

$$u_0(x) = f_0(x), \tag{40}$$

$$u_1(x) = f_1(x) + \lambda \int_0^x K(x,t)u_0(t)dt, \tag{41}$$

$$u_2(x) \;=\; \lambda \int_0^x K(x,t)u_1(t)dt, \tag{42}$$

$$u_3(x) \;=\; \lambda \int_0^x K(x,t)u_2(t)dt, \tag{43}$$

and so on. The method discussed above for the determination of the components of the solution $u(x)$ of Eq. (1) can be written in a recursive relationship by

$$u_0(x) \;=\; f_0(x), \tag{44}$$

$$u_1(x) \;=\; f_1(x) + \lambda \int_0^x K(x,t)u_0(t)dt, \tag{45}$$

$$u_{n+1}(x) \;=\; \lambda \int_0^x K(x,t)u_n(t)dt, \; n \geq 1. \tag{46}$$

In most problems, we need to use (44) and (45) only. For illustration purposes, we study the following examples.

Example 5. We consider here the Volterra integral equation

$$u(x) = \sec x \tan x - \frac{1}{4}\left(e^{\sec x} - e\right)x + \frac{1}{4}\int_0^x x e^{\sec t} u(t)\, dt, \; x < \frac{\pi}{2}. \tag{47}$$

Using the modified decomposition method as discussed above, we first decompose the function $f(x)$ into

$$f_0(x) = \sec x \tan x, \tag{48}$$

and

$$f_1(x) = -\frac{1}{4}\left(e^{\sec x} - e\right)x. \tag{49}$$

Consequently, we find

$$u_0(x) \;=\; \sec x \tan x,$$

$$u_1(x) \;=\; -\frac{1}{4}\left(e^{\sec x} - e\right)x + \frac{1}{4}\int_0^x x e^{\sec t} u_0(t)\, dt$$

$$\;=\; -\frac{1}{4}\left(e^{\sec x} - e\right)x + \frac{1}{4}x\int_0^x \sec t\,\tan t\, e^{\sec t}\, dt = 0,$$

obtained by integrating by substitution where we set $y = \sec t$. Accordingly, other components $u_i(x) = 0$, for $i \geq 2$. Therefore, the exact solution

$$u(x) = \sec x \tan x, \tag{50}$$

follows immediately. It is clear that two components are calculated to determine the exact solution. modified recurrence relation.

Example 6. We consider here the Volterra integral equation

$$u(x) = \cos x + \sin x - \int_0^x u(t)\, dt. \tag{51}$$

Using the modified decomposition method as discussed above, we first decompose the function $f(x)$ into

$$f_0(x) = \cos x, \tag{52}$$

and

$$f_1(x) = \sin x. \tag{53}$$

Consequently, we find

$$u_0(x) = \cos x,$$

$$u_1(x) = -\int_0^x u_0(t)\, dt = 0.$$

Accordingly, other components $u_i(x) = 0$, for $i \geq 2$. Therefore, the exact solution

$$u(x) = \cos x, \tag{54}$$

follows immediately.

3.2.2 The Noise Terms Phenomenon

The noise terms phenomenon was presented before in studying the Fredholm integral equations. It was proved that the noise terms phenomenon accelerates the convergence of the solution. The noise terms, that may appear between components of the unknown solution, are defined by the identical terms with opposite signs. The noise terms, that may exist between the first two components $u_0(x)$ and $u_1(x)$, may provide the exact solution by using these two components.

In Chapter 2, we outlined the main concepts of the noise terms phenomenon, hence we present a summary of these outlines. When the noise terms appear, especially between the components $u_0(x)$ $u_1(x)$, then by canceling the noise terms between $u_0(x)$ and $u_1(x)$, even though $u_1(x)$ contains further terms, the remaining non-canceled terms of $u_0(x)$ may give the exact solution of the integral equation. The appearance of the noise terms between $u_0(x)$ and $u_1(x)$ is not always enough to give the exact solution, hence it is necessary to show that the non-canceled terms of $u_0(x)$ satisfy the given integral equation.

We point out that noise terms may appear only for nonhomogeneous problems, whereas homogeneous problems do not give rise to noise terms. Moreover, it was proved by Wazwaz in [62] that the appearance of the noise terms is governed by a necessary condition, in that the zeroth component $u_0(x)$ must contain the exact solution $u(x)$ among other terms. The phenomenon of the useful noise terms will be explained by the following illustrative examples.

Example 7. Solve the Volterra integral equation by using noise terms phenomenon

$$u(x) = 6x + 3x^2 - \int_0^x u(t)\, dt. \tag{55}$$

The Adomian method admits the use of the recurrence relation

$$\begin{aligned} u_0(x) &= 6x + 3x^2, \\ u_{k+1}(x) &= -\int_0^x t u_k(t)\, dt, \ k \geq 0. \end{aligned} \tag{56}$$

This gives

$$\begin{aligned} u_0(x) &= 6x + 3x^2, \\ u_1(x) &= -\int_0^x t u_0(t)\, dt = -3x^2 - x^3. \end{aligned} \tag{57}$$

The noise terms $\pm 3x^2$ appear in $u_0(x)$ and $u_1(x)$. Canceling this term from the zeroth component $u_0(x)$ gives the exact solution

$$u(x) = 6x, \tag{58}$$

that satisfies the integral equation. This should be justified by substitution.

Example 8. Solve the Volterra integral equation by using noise terms phenomenon

$$u(x) = \sinh x + x \sinh x - x^2 \cosh x + \int_0^x x t u(t)\, dt. \tag{59}$$

Following the standard Adomian method we set the recurrence relation

$$\begin{aligned} u_0(x) &= \sinh x + x \sinh x - x^2 \cosh x, \\ u_{k+1}(x) &= \int_0^x x t u_k(t)\, dt, \ k \geq 0. \end{aligned} \tag{60}$$

This gives

$$\begin{aligned} u_0(x) &= \sinh x + x \sinh x - x^2 \cosh x, \\ u_1(x) &= \int_0^x t u_0(t)\, dt = -x \sinh x + x^2 \cosh x + \text{other terms}. \end{aligned} \tag{61}$$

The noise terms $\pm x \sinh x$ and $\mp x^2 \cosh x$ appear in $u_0(x)$ and $u_1(x)$. Canceling these terms from the zeroth component $u_0(x)$ gives the exact solution

$$u(x) = \sinh x, \tag{62}$$

that satisfies the integral equation, that should be justified. The other terms of $u_1(x)$ vanish in the limit with other terms of the other components.

Exercises 3.2

Solve the following Volterra integral equations by using the *Adomian decomposition method*

1. $u(x) = 4x + 2x^2 - \displaystyle\int_0^x u(t)dt$

2. $u(x) = 1 + x - x^2 + \displaystyle\int_0^x u(t)dt$

3. $u(x) = 1 - \displaystyle\int_0^x u(t)dt$

4. $u(x) = x + \displaystyle\int_0^x (x - t)u(t)dt$

5. $u(x) = 3x - 9 \displaystyle\int_0^x (x - t)u(t)dt$

6. $u(x) = 1 - 4 \displaystyle\int_0^x (x - t)u(t)dt$

7. $u(x) = 1 + x - \displaystyle\int_0^x (x - t)u(t)dt$

8. $u(x) = 1 - x - \displaystyle\int_0^x (x - t)u(t)dt$

9. $u(x) = 1 + x + \displaystyle\int_0^x (x - t)u(t)dt$

10. $u(x) = 1 - x + \displaystyle\int_0^x (x - t)u(t)dt$

11. $u(x) = 2 + \displaystyle\int_0^x (x - t)u(t)dt$

12. $u(x) = 1 + x + \displaystyle\int_0^x u(t)dt$

13. $u(x) = 1 - \dfrac{1}{2!}x^2 - \displaystyle\int_0^x (x - t)u(t)dt$

14. $u(x) = 1 + \dfrac{1}{2!}x^2 + \displaystyle\int_0^x (x - t)u(t)dt$

In exercises 15-19 solve the given Volterra integral equations by using the *modified decomposition method*:

15. $u(x) = \cos x + \left(1 - e^{\sin x}\right) x + x \int_0^x e^{\sin t} u(t) dt$

16. $u(x) = \sec^2 x + \left(1 - e^{\tan x}\right) x + x \int_0^x e^{\tan t} u(t) dt, \ x < \pi/2$

17. $u(x) = \cosh x + \dfrac{x}{2} \left(1 - e^{\sinh x}\right) + \dfrac{x}{2} \int_0^x e^{\sinh t} u(t) dt$

18. $u(x) = \sinh x + \dfrac{1}{10} \left(e - e^{\cosh x}\right) + \dfrac{1}{10} \int_0^x e^{\cosh t} u(t) dt$

19. $u(x) = x^3 - x^5 + 5 \int_0^x t u(t) dt$

20. $u(x) = \sec x \tan x + (e - e^{\sec x}) + \int_0^x e^{\sec t} u(t) dt, \ x < \pi/2$

In exercises 21-26 solve the given Volterra integral equations by using the *noise terms phenomenon:*

21. $u(x) = 8x - 4x^3 + \int_0^x x u(t) dt$

22. $u(x) = 8x^2 - 2x^5 + \int_0^x x t u(t) dt$

23. $u(x) = \sec^2 x - \tan x + \int_0^x u(t) dt$

24. $u(x) = 1 + x + x^2 - \dfrac{1}{4}x^4 - \dfrac{1}{5}x^5 - \dfrac{1}{6}x^6 + \int_0^x t^3 u(t) dt$

25. $u(x) = x \sin x + x \cos x - \sin x + \int_0^x u(t) dt$

26. $u(x) = \cosh^2 x - \dfrac{1}{4} \sinh(2x) - \dfrac{1}{2}x + \int_0^x u(t) dt$

3.3   The Variational Iteration Method

In Chapter 2, we presented the *variational iteration method* for handling the Fredholm integral equations. Ji-Huan He [15–17] developed this method, that proved to be reliable in the study of linear and nonlinear, and homogeneous and inhomogeneous equations. As we showed in Chapter 2, the method gives the successive approximations of the exact solution that may converge to the exact solution in case this solution exists. One significant feature of this method is that it can handle especially nonlinear problems without the use of the so-called Adomian polynomials as required by the Adomian decomposition method.

     The variational iteration method was given in detail in Chapter 2, hence we skip it here. The method admits the use of a correction functional in the form

$$u_{n+1}(x) = u_n(x) + \int_0^x \lambda(\xi)\left(Lu_n(\xi) + N\,\tilde{u}_n(\xi) - g(\xi)\right)d\xi, n \geq 0, \quad (63)$$

where λ is a general Lagrange multiplier that can be determined optimally via the variational theory as shown before. In Chapter 2, we presented a rule that gives these Lagrange multipliers for some ordinary differential equations that will be examined in this text. Recall that the variational iteration method (VIM) is used for ODEs and integro-differential equations and this can be obtained by differentiating the Volterra integral equation. Moreover, the zeroth component $u_0(x)$ in (63) can be selected according to the order of the resulted ODE. For example, for ODEs of first order, second order, and third order, we select

$$\begin{aligned} u_0(x) &= u(0), \\ u_0(x) &= u(0) + xu^{'}(0), \\ u_0(x) &= u(0) + xu^{'}(0) + \tfrac{1}{2!}x^2 u^{''}(0), \end{aligned} \qquad (64)$$

respectively, and these are the first term, the first two terms, and the first three terms of the Taylor series of $u(x)$ at $x = 0$. The exact solution is thus given by

$$u(x) = \lim_{n \to \infty} u_n(x). \qquad (65)$$

In other words, to solve any Volterra integral equation by using this method, we should first transform this equation to its equivalent ODE or its equivalent Volterra integro-differential equation, where Leibniz rule should be used. The next step consists of the determination of the Lagrange multiplier λ using the rules given in Chapter 2. Finally, we select the zeroth approximation $u_0(x)$ as indicated earlier. Having prepared all these steps, we then use the correction functional (63) to determine as many successive approximations as we can.

Example 1. Solve the Volterra integral equation by using the variational iteration method

$$u(x) = 1 - x + \int_0^x (x - t)u(t)\, dt. \qquad (66)$$

Differentiating both sides of (66), and using Leibniz rule, we find

$$u^{'}(x) = -1 + \int_0^x u(t)\, dt. \qquad (67)$$

The initial condition $u(0) = 1$ is obtained by using $x = 0$ into (66), and hence we can select $u_0(x) = 1$.

The correction functional for equation (67) is

$$u_{n+1}(x) = u_n(x) - \int_0^x \left(u_n'(\xi) + 1 - \int_0^\xi u_n(r)\, dr \right) d\xi, \qquad (68)$$

where we selected $\lambda = -1$ for the first order integro-differential equation (67). As stated before, we can use the initial condition to select $u_0(x) = 1$ that will lead to the following successive approximations

$$
\begin{aligned}
u_0(x) &= 1, \\
u_1(x) &= 1 - \int_0^x \left(u_0'(\xi) + 1 - \int_0^\xi u_0(r)\, dr \right) d\xi = 1 - x + \frac{1}{2!}x^2, \\
u_2(x) &= 1 - x + \frac{1}{2!}x^2 - \int_0^x \left(u_1'(\xi) - \int_0^\xi u_1(r)\, dr \right) d\xi \\
&= 1 - x + \frac{1}{2!}x^2 - \frac{1}{3!}x^3 + \frac{1}{4!}x^4.
\end{aligned}
$$

$$\qquad (69)$$

Recall that

$$
\begin{aligned}
u(x) &= \lim_{n\to\infty} u_n(x), \\
&= \lim_{n\to\infty} \left(1 - x + \frac{1}{2!}x^2 - \frac{1}{3!}x^3 + \frac{1}{4!}x^4 - \cdots + (-1)^n \frac{1}{n!}x^n \right),
\end{aligned}
\qquad (70)
$$

that gives the exact solution by

$$u(x) = e^{-x}. \qquad (71)$$

Example 2. Solve the Volterra integral equation by using the variational iteration method

$$u(x) = 2\sin x - \frac{1}{6}x^3 + \int_0^x (x - t)\, u(t)\, dt. \qquad (72)$$

Differentiating both sides of (72) once with respect to x and using Leibniz rule we obtain the integro-differential equation

$$u'(x) = 2\cos x - \frac{1}{2}x^2 + \int_0^x u(t)\, dt, \qquad (73)$$

where by using $x = 0$ into (72) we find $u(0) = 0$.

The variational iteration method admits the use of correction functional for equation (73) by

$$u_{n+1}(x) = u_n(x) - \int_0^x \left(u_n'(\xi) - 2\cos(\xi) + \frac{1}{2}\xi^2 - \int_0^\xi u_n(r)\, dr \right) d\xi. \quad (74)$$

Proceeding as before we obtain the following successive approximations

$$
\begin{aligned}
u_0(x) &= 0, \\
u_1(x) &= -\int_0^x \left(u_0'(\xi) - 2\cos(\xi) + \frac{1}{2}\xi^2 - \int_0^\xi u_0(r)\,dr \right) d\xi \\
&= 2\sin x - \frac{1}{6}x^3, \\
u_2(x) &= 2\sin x - \frac{1}{6}x^3 \\
&\quad - \int_0^x \left(u_1'(\xi) - 2\cos(\xi) + \frac{1}{2}\xi^2 - \int_0^\xi u_1(r)\,dr \right) d\xi \\
&= 2x - \frac{1}{6}x^3 - \frac{1}{120}x^5, \\
u_3(x) &= 2x - \frac{1}{3!}x^3 + \frac{1}{5!}x^5 - \frac{1}{7!}x^7,
\end{aligned}
\tag{75}
$$

and so on. The solution in a series form is given by

$$
u(x) = x + \left(x - \frac{1}{3!}x^3 + \frac{1}{5!}x^5 - \frac{1}{7!}x^7 + \cdots \right).
\tag{76}
$$

Recall that

$$
u(x) = \lim_{n \to \infty} u_n(x),
\tag{77}
$$

and this gives the exact solution by

$$
u(x) = x + \sin x.
\tag{78}
$$

Example 3. Solve the Volterra integral equation by using the variational iteration method

$$
u(x) = 3e^x - 3 - 2x - \frac{1}{2}x^2 + \frac{1}{2}\int_0^x (x-t)^2\, u(t)\, dt.
\tag{79}
$$

Using Leibniz rule to differentiate both sides of (79) with respect to x gives the integro-differential equations

$$
u'(x) = 1 + x + \int_0^x (x-t)u(t)dt, \quad u(0) = 0.
\tag{80}
$$

Proceeding as before we obtain the following successive approximations

$$
\begin{aligned}
u_0(x) &= 0, \\
u_1(x) &= x + x^2 + \tfrac{1}{2}x^3 + \tfrac{1}{8}x^4 + \tfrac{1}{40}x^5 + \cdots, \\
u_2(x) &= x + x^2 + \tfrac{1}{2!}x^3 + \tfrac{1}{3!}x^4 + \tfrac{1}{4!}x^5 + \cdots, \\
u_n(x) &= x(1 + x + \tfrac{1}{2!}x^2 + \tfrac{1}{3!}x^3 + \tfrac{1}{4!}x^4 + \tfrac{1}{5!}x^5 + \tfrac{1}{6!}x^6 + \cdots + \tfrac{1}{n!}x^n).
\end{aligned}
\tag{81}
$$

This in turn gives the exact solution by

$$u(x) = xe^x. \tag{82}$$

Exercises 3.3

Use the *variational iteration method* to solve the following Volterra integral equations:

1. $u(x) = 1 + \int_0^x u(t)dt$

2. $u(x) = x + \int_0^x (x - t)u(t)dt$

3. $u(x) = 3x - 9 \int_0^x (x - t)u(t)dt$

4. $u(x) = 1 - 4 \int_0^x (x - t)u(t)dt$

5. $u(x) = 1 + x - \int_0^x (x - t)u(t)dt$

6. $u(x) = 1 - x - \int_0^x (x - t)u(t)dt$

7. $u(x) = 1 - \frac{1}{2}x^2 + \sin x - \cos x + \frac{1}{2} \int_0^x (x - t)^2 u(t)dt$

8. $u(x) = x + \cosh x - \sinh x + \frac{1}{2} \int_0^x (x - t)^2 u(t)dt$

3.4   The Series Solution Method

In this section we will introduce a practical method to solve the Volterra integral equation with variable limits of integration

$$u(x) = f(x) + \lambda \int_0^x K(x, t)u(t)dt, \tag{83}$$

where $K(x, t)$ is the kernel of the integral equation, and λ is a parameter, called the *series solution method*. We will follow a parallel approach to the method of the series solution around an ordinary point that usually applied in solving an ordinary differential equation. The method is applicable provided that $u(x)$ is an analytic function, i.e. $u(x)$ has a Taylor expansion at $x = 0$. Accordingly, $u(x)$ can be expressed by a series expansion given by

$$u(x) = \sum_{n=0}^{\infty} a_n x^n, \tag{84}$$

where the coefficients a_n are constants that will be determined. Substituting (84) into both sides of (83) yields

$$\sum_{n=0}^{\infty} a_n x^n = f(x) + \lambda \int_0^x K(x,t) \left(\sum_{n=0}^{\infty} a_n t^n \right) dt, \qquad (85)$$

so that by using few terms of the expansions in both sides, we find

$$a_0 + a_1 x + a_2 x^2 + a_3 x^3 + \cdots = f(x) + \lambda \int_0^x K(x,t) a_0 \, dt$$

$$+ \lambda \int_0^x K(x,t) a_1 \, t \, dt$$

$$+ \lambda \int_0^x K(x,t) a_2 \, t^2 \, dt \qquad (86)$$

$$+ \lambda \int_0^x K(x,t) a_3 \, t^3 \, dt$$

$$+ \cdots.$$

In view of (86), the integral equation (83) will be reduced to several traditional integrals, with defined integrands having terms of the form t^n, $n \geq 0$ only. We then write the Taylor expansions for $f(x)$ and evaluate the first few integrals in (86). Having performed the evaluation, we equate the coefficients of like powers of x in both sides of Eq. (86). This will lead to a complete determination of the coefficients $a_0, a_1, a_2, \ldots$. Consequently, substituting the obtained coefficients a_n, $n \geq 0$, in (84) produces the solution in a series form. This may lead to a solution in a closed form if the expansion obtained is a Taylor expansion to a well known elementary function.

It seems reasonable to illustrate the series solution method by discussing the following examples.

Example 1. We use the series solution method to solve

$$u(x) = 1 + \int_0^x (t - x) u(t) dt. \qquad (87)$$

Substituting $u(x)$ by the series

$$u(x) = \sum_{n=0}^{\infty} a_n x^n, \qquad (88)$$

into both sides of the equation (87) leads to

$$\sum_{n=0}^{\infty} a_n x^n = 1 + \int_0^x (t - x) \left(\sum_{n=0}^{\infty} a_n t^n \right) dt, \qquad (89)$$

which gives

$$\sum_{n=0}^{\infty} a_n x^n = 1 + \int_0^x \left(\sum_{n=0}^{\infty} a_n t^{n+1} - x \sum_{n=0}^{\infty} a_n t^n \right) dt. \tag{90}$$

Evaluating the regular integrals at the right hand side that involve terms of the form t^n, $n \geq 0$ yields

$$\sum_{n=0}^{\infty} a_n x^n = 1 - \sum_{n=0}^{\infty} \frac{1}{(n+1)(n+2)} a_n x^{n+2}, \tag{91}$$

or equivalently

$$a_0 + a_1 x + a_2 x^2 + a_3 x^3 + \cdots = 1 - \frac{1}{2!} a_0 x^2 - \frac{1}{3!} a_1 x^3 - \frac{1}{12} a_2 x^4 + \cdots \tag{92}$$

Equating the coefficients of like powers of x in both sides we find

$$\begin{array}{rcl}
a_0 &=& 1, \\
a_1 &=& 0, \\
a_2 &=& -\frac{1}{2!}, \\
a_3 &=& 0, \\
a_4 &=& \frac{1}{4!},
\end{array} \tag{93}$$

and generally

$$a_{2n} = (-1)^n \frac{1}{(2n)!}, \quad \text{for} \quad n \geq 0, \tag{94}$$

$$a_{2n+1} = 0, \quad \text{for} \quad n \geq 0. \tag{95}$$

Using (88) we find the solution in a series form

$$u(x) = 1 - \frac{1}{2!} x^2 + \frac{1}{4!} x^4 - \frac{1}{6!} x^6 + \dots, \tag{96}$$

and in a closed form by

$$u(x) = \cos x. \tag{97}$$

Example 2. We next use the series solution method to solve

$$u(x) = 2 \cosh x - x \sinh x - 1 + \int_0^x t u(t) dt. \tag{98}$$

Proceeding as before, and substituting (88) into both sides of the equation (98) we obtain

$$\sum_{n=0}^{\infty} a_n x^n = 2 \left(\sum_{n=0}^{\infty} \frac{x^{2n}}{(2n)!} \right) - x \left(\sum_{n=0}^{\infty} \frac{x^{2n+1}}{(2n+1)!} \right) - 1 + \int_0^x t \left(\sum_{n=0}^{\infty} a_n t^n \right) dt, \tag{99}$$

where we use the Taylor series of $\cosh x$ and $\sinh x$. Using a few terms from each series involved and integrating the resulting integrals at the right hand side we obtain

$$a_0 + a_1 x + a_2 x^2 + \cdots = 2\left(1 + \frac{x^2}{2!} + \frac{x^4}{4!} + \cdots\right) - x\left(x + \frac{x^3}{3!} + \frac{x^5}{5!} \cdots\right)$$

$$-1 + \frac{1}{2}a_0 x^2 + \frac{1}{3}a_1 x^3 + \frac{1}{4}a_2 x^4 + \cdots \tag{100}$$

Equating the coefficients of like powers of x in in both sides of (100) yields

$$\begin{aligned}
a_0 &= 1, \\
a_1 &= 0, \\
a_2 &= \frac{1}{2!}, \\
a_3 &= 0, \\
a_4 &= \frac{1}{4!},
\end{aligned} \tag{101}$$

and generally

$$a_{2n} = \frac{1}{(2n)!}, \quad \text{for } n \geq 0, \tag{102}$$

$$a_{2n+1} = 0, \quad \text{for } n \geq 0. \tag{103}$$

Consequently, the solution in a series form is given by

$$u(x) = 1 + \frac{1}{2!}x^2 + \frac{1}{4!}x^4 + \frac{1}{6!}x^6 + \cdots, \tag{104}$$

and in a closed form

$$u(x) = \cosh x, \tag{105}$$

obtained by using the Taylor expansion of $\cosh x$.

Exercises 3.4

Solve the following Volterra integral equations by using the *series solution method*

1. $u(x) = 2x + 2x^2 - x^3 + \displaystyle\int_0^x u(t)dt$

2. $u(x) = 1 + x - \dfrac{2}{3}x^3 - \dfrac{1}{2}x^4 + 2\displaystyle\int_0^x tu(t)dt$

3. $u(x) = 1 + 2\sin x - \displaystyle\int_0^x u(t)dt$

4. $u(x) = 1 + x + \dfrac{1}{2!}x^2 + \dfrac{1}{3!}x^3 - \displaystyle\int_0^x (x-t)u(t)dt$

5. $u(x) = -1 - \displaystyle\int_0^x u(t)dt$

6. $u(x) = 1 - 2\displaystyle\int_0^x u(t)dt$

7. $u(x) = 1 + xe^x - \displaystyle\int_0^x tu(t)dt$

8. $u(x) = x + \displaystyle\int_0^x (x - t)u(t)dt$

9. $u(x) = 1 - \dfrac{1}{2!}x^2 - \displaystyle\int_0^x (x - t)u(t)dt$

10. $u(x) = 1 - x - \displaystyle\int_0^x (x - t)u(t)dt$

11. $u(x) = 1 + \sinh x - \cosh x + \displaystyle\int_0^x u(t)dt$

12. $u(x) = x\cos x + \displaystyle\int_0^x tu(t)dt$

3.5   Converting Volterra Equation to IVP

In Chapter 1 the process of converting initial value problems to equivalent Volterra integral equations has been discussed in detail. However, the technique of converting Volterra integral equations to initial value problems will be discussed in this section, though it is rarely used. This may be explained by the fact that integral equations are easily solved because initial conditions are embedded in the integral equations. However, solving initial value problems, where initial conditions will be used, will increase the size of evaluations required because additional steps will be needed to complete the solution.

To employ this method, we simply differentiate both sides of (1), noting that Leibniz rule should be used in differentiating the integral at the right hand side of (1). The differentiation process should be carried out successively until the integral sign is removed and the integral equation is converted to a pure differential equation equivalent to the integral equation under discussion. It is interesting to note that initial conditions should be determined at every step of differentiating by setting $x = 0$ at $u(x)$ and its obtained derivatives. The resulting initial value problem is then solved following the traditional techniques used in undergraduate course of ordinary differential equations. The technique of converting Volterra integral equations to initial value problems, though not usually used as indicated earlier, will be illustrated by discussing the following examples.

Example 1. Solve the following Volterra integral equation

$$u(x) = x^2 + \frac{1}{12}x^4 + \int_0^x (t-x)\, u(t)\, dt, \qquad (106)$$

by converting it to an equivalent initial value problem.

Differentiating both sides of (106) with respect to x and using the Leibniz rule we find

$$u'(x) = 2x + \frac{1}{3}x^3 - \int_0^x u(t)\, dt. \qquad (107)$$

Note that we have to differentiate both sides of (107) again to get rid of the integral sign at the right hand side, doing so we obtain

$$u''(x) = 2 + x^2 - u(x), \qquad (108)$$

or equivalently the nonhomogeneous second order differential equation

$$u''(x) + u(x) = 2 + x^2. \qquad (109)$$

The proper initial conditions can be obtained by substituting $x = 0$ into both sides of equations (106) and (107), hence we find

$$u(0) = 0, \quad u'(0) = 0. \qquad (110)$$

To solve the resulting initial value problem

$$u''(x) + u(x) = 2 + x^2, \quad u(0) = 0, \quad u'(0) = 0. \qquad (111)$$

We first solve the corresponding homogeneous equation

$$u''(x) + u(x) = 0. \qquad (112)$$

The characteristic equation of (112) is given by

$$r^2 + 1 = 0, \qquad (113)$$

so that the roots of (113) are

$$r_1 = i, \quad r_2 = -i. \qquad (114)$$

As a result the complementary solution is

$$u_c = A\cos x + B\sin x, \qquad (115)$$

where A and B are constants to be determined later by using the initial conditions. To determine a particular solution $u_p(x)$ of (111), we assume that $u_p(x)$ is of the form

$$u_p = \alpha + \beta x + \gamma x^2, \tag{116}$$

where α, β, and γ are constants that will be determined. Substituting (116) into (111), and equating like powers of x on each side we find

$$\alpha = 0, \quad \beta = 0, \quad \gamma = 1. \tag{117}$$

Combining (115) – (117) we obtain the general solution

$$u(x) = A \cos x + B \sin x + x^2. \tag{118}$$

We can determine the constants A and B upon using the initial conditions (110) where we find

$$A = 0, \quad B = 0, \tag{119}$$

so that the solution of (111)

$$u(x) = x^2, \tag{120}$$

follows immediately.

However, we can easily show that using the modified decomposition method, by setting $u_0(x) = x^2$, will give the same result obtained above with minimal work.

Example 2. Solve the following Volterra integral equation

$$u(x) = e^x + \int_0^x (t - x)\, u(t)\, dt, \tag{121}$$

by converting to an equivalent initial value problem.

Differentiating both sides of (121) twice with respect to x and using the Leibniz rule we obtain

$$u'(x) = e^x - \int_0^x u(t)\, dt, \tag{122}$$

and

$$u''(x) = e^x - u(x) \tag{123}$$

or equivalently

$$u''(x) + u(x) = e^x. \tag{124}$$

The proper initial conditions can be obtained by substituting $x = 0$ into both sides of the equations (121) and (122), hence we find

$$u(0) = 1, \quad u^{'}(0) = 1. \tag{125}$$

To solve the resulting initial value problem

$$u^{''}(x) + u(x) = e^x, \quad u(0) = 1, \quad u^{'}(0) = 1. \tag{126}$$

We first solve the corresponding homogeneous equation

$$u^{''}(x) + u(x) = 0. \tag{127}$$

Following the steps we used in the first example, we find that the complementary solution is

$$u_c = A\cos x + B\sin x, \tag{128}$$

where A and B are constants to be determined later by using the initial conditions. Moreover, a particular solution of (126) may be assumed of the form

$$u_p = \alpha e^x, \tag{129}$$

where α is a constant that will be determined. Substituting (129) into (126) and equating like powers of x on both sides we find

$$\alpha = \frac{1}{2}. \tag{130}$$

Combining (128) – (130) we obtain the general solution

$$u(x) = A\cos x + B\sin x + \frac{1}{2}e^x. \tag{131}$$

The constants A and B are determined by using the initial conditions in (125) where we find

$$A = \frac{1}{2}, \quad B = \frac{1}{2}. \tag{132}$$

Thus the solution of (126) is

$$u(x) = \frac{1}{2}\left(\sin x + \cos x + e^x\right), \tag{133}$$

obtained upon substituting (132) into (131).

Exercises 3.5

Solve the following Volterra integral equations by converting to equivalent *initial value problems:*

1. $u(x) = 1 - 3 \int_0^x u(t)dt$

2. $u(x) = 1 + \int_0^x (x-t)u(t)dt$

3. $u(x) = 1 - x - \int_0^x (x-t)u(t)dt$

4. $u(x) = x + \int_0^x u(t)dt$

5. $u(x) = 1 + x + \int_0^x (x-t)u(t)dt$

6. $u(x) = 1 + \dfrac{1}{6} \int_0^x (x-t)^3 u(t)dt$

7. $u(x) = x + \dfrac{1}{6} \int_0^x (x-t)^3 u(t)dt$

8. $u(x) = x^2 + \int_0^x (x-t)u(t)dt$

9. $u(x) = x + \dfrac{1}{3!}x^3 - \int_0^x (x-t)u(t)dt$

10. $u(x) = x - x^2 + \dfrac{1}{6}x^3 - \dfrac{1}{12}x^4 - \int_0^x (x-t)u(t)dt$

3.6   Successive Approximations Method

The method of successive approximations, that was used earlier in Chapter 2 for handling Fredholm integral equation will be implemented here to solve Volterra integral equations as well. In this method, we replace the unknown function $u(x)$ under the integral sign of the Volterra equation

$$u(x) = f(x) + \lambda \int_0^x K(x,t)u(t)dt, \tag{134}$$

by any selective real valued continuous function $u_0(x)$, called the zeroth approximation. This substitution will give the first approximation $u_1(x)$ by

$$u_1(x) = f(x) + \lambda \int_0^x K(x,t)u_0(t)dt. \tag{135}$$

It is obvious that $u_1(x)$ is continuous whenever $f(x)$, $K(x,t)$ and $u_0(x)$ are continuous. The second approximation $u_2(x)$ of $u(x)$ can be obtained similarly by replacing $u_0(x)$ in (135) by $u_1(x)$ obtained above, hence we find

$$u_2(x) = f(x) + \lambda \int_0^x K(x,t)u_1(t)dt. \tag{136}$$

This process can be continued in the same manner to obtain the nth approximation. In other words, the various approximations of the solution $u(x)$ of (134) can be obtained in a recursive scheme given by

$$
\begin{cases}
u_0(x) & = \quad \text{any selective real valued function} \\
u_n(x) & = \quad f(x) + \lambda \displaystyle\int_0^x K(x,t)u_{n-1}(t)dt, \quad n \geq 1.
\end{cases} \tag{137}
$$

The most commonly selected functions for $u_0(x)$ are 0, 1 or x. At the limit, the solution $u(x)$ of the equation (134) is obtained by

$$
u(x) = \lim_{n \to \infty} u_n(x), \tag{138}
$$

so that the resulting solution $u(x)$ is independent of the choice of the zeroth approximation $u_0(x)$.

It is useful, for comparison reasons, to distinguish between the recursive schemes used in the decomposition method and in the successive approximations method. In the decomposition method, we decompose the solution $u(x)$ into components $u_0, u_1, u_2, \dots$ where each component is evaluated subsequently, and in this case the solution is given in a series form

$$
u(x) = \sum_{n=0}^{\infty} u_n(x), \tag{139}
$$

where the zeroth component $u_0(x)$ is defined and given by all terms that are out of the integral sign. However, in the successive approximations method, we apply the above recursive scheme (137) to determine various approximations of the solution $u(x)$ itself, and not components of $u(x)$, noting that the zeroth approximation is not defined but rather given by a selective real valued function, and as a result the solution $u(x)$ is given by the formula (138).

To illustrate the difference between the two recursive algorithms, we start by solving Example 2 in Section 3.2. The reader can easily compare between the two approaches.

Example 1. Solve the Volterra integral equation

$$
u(x) = x + \int_0^x (t - x)u(t)\, dt, \tag{140}
$$

by the *successive approximations method*. We first select any real valued function for the zeroth approximation, hence we set

$$
u_0(x) = 0. \tag{141}
$$

Substituting (141) into (140) we find

$$u_1(x) = x + \int_0^x (t - x)\, u_0(t)\, dt, \qquad (142)$$

and this gives the first approximation of $u(x)$ by

$$u_1(x) = x. \qquad (143)$$

Inserting (143) into (142) to replace $u_0(x)$ we obtain

$$u_2(x) = x + \int_0^x (t - x)\, t\, dt, \qquad (144)$$

where by integrating the right hand side of (144), the second approximation of $u(x)$

$$u_2(x) = x - \frac{1}{3!}x^3, \qquad (145)$$

is readily obtained. Continuing in the same manner we find that the third approximation of $u(x)$ is

$$u_3(x) = x - \frac{1}{3!}x^3 + \frac{1}{5!}x^5. \qquad (146)$$

Accordingly, the nth approximation is given by

$$u_n(x) = \sum_{k=1}^{n} (-1)^{k-1} \frac{x^{2k-1}}{(2k-1)!}, \quad n \geq 1. \qquad (147)$$

Consequently, the solution $u(x)$ of (140) is given by

$$
\begin{aligned}
u(x) &= \lim_{n \to \infty} u_n(x) \\
&= \lim_{n \to \infty} \left(\sum_{k=1}^{n} (-1)^{k-1} \frac{x^{2k-1}}{(2k-1)!} \right) \qquad (148) \\
&= \sin x.
\end{aligned}
$$

To show that $u(x)$ obtained in (148) does not depend on the selection of $u_0(x)$, we will solve the equation (140) by selecting

$$u_0(x) = x. \qquad (149)$$

Using the new selection of $u_0(x)$ in the right hand side of (140) we obtain

$$u_1(x) = x + \int_0^x (t - x)\, t\, dt, \qquad (150)$$

which gives the first approximation by

$$u_1(x) = x - \frac{1}{3!}x^3. \tag{151}$$

Proceeding as before we can easily obtain the second approximation

$$u_2(x) = x - \frac{1}{3!}x^3 + \frac{1}{5!}x^5. \tag{152}$$

In a parallel manner we find that

$$u_n(x) = \sum_{k=0}^{n} (-1)^k \frac{x^{2k+1}}{(2k+1)!}, \ n \geq 0. \tag{153}$$

Accordingly, we obtain

$$\begin{aligned} u(x) &= \lim_{n\to\infty} u_n(x) \\ &= \lim_{n\to\infty} \left(\sum_{k=0}^{n} (-1)^k \frac{x^{2k+1}}{(2k+1)!} \right) \\ &= \sin x, \end{aligned} \tag{154}$$

the same answer we obtained above in (148). This confirms the fact that the solution obtained does not depend on the selection of the zeroth approximation $u_0(x)$.

Example 2. Solve the Volterra integral equation

$$u(x) = 1 - \int_0^x (t - x)\, u(t)dt, \tag{155}$$

by using the successive approximations method. We start first by selecting the zeroth approximation and in this time we choose

$$u_0(x) = 1, \tag{156}$$

where by substituting this in the right hand side of (155) the first approximation

$$u_1(x) = 1 - \int_0^x (t - x)u_0(t)\, dt, \tag{157}$$

so that

$$u_1(x) = 1 + \frac{1}{2!}x^2 \tag{158}$$

follows immediately. Proceeding in the same manner we find that

$$u_2(x) = 1 - \int_0^x (t - x)\, u_1(t)dt, \tag{159}$$

so that

$$u_2(x) = 1 + \frac{1}{2!}x^2 + \frac{1}{4!}x^4. \tag{160}$$

In a similar manner we obtain

$$u_3(x) = 1 + \frac{1}{2!}x^2 + \frac{1}{4!}x^4 + \frac{1}{6!}x^4. \tag{161}$$

Generally we obtain for the nth approximation

$$u_n(x) = \sum_{k=0}^{n} \frac{x^{2k}}{(2k)!}, \quad n \geq 0. \tag{162}$$

Consequently, the solution $u(x)$ of (155) is given by

$$\begin{aligned}
u(x) &= \lim_{n \to \infty} u_n(x), \\
&= \lim_{n \to \infty} \left(\sum_{k=0}^{n} \frac{x^{2k}}{(2k)!} \right) \\
&= \cosh x,
\end{aligned} \tag{163}$$

obtained upon using the Taylor expansion of $\cosh x$.

It is useful to observe that the zeroth approximation in this method is selected and it is not a part of the integral equation as in the decomposition method.

Exercises 3.6

Solve the following Volterra integral equations by the *successive approximations method*:

1. $u(x) = 1 - \displaystyle\int_0^x u(t)dt$

2. $u(x) = 1 - 9 \displaystyle\int_0^x (x - t)u(t)dt$

3. $u(x) = 1 + 2x + 4 \displaystyle\int_0^x (x - t)u(t)dt$

4. $u(x) = 1 - \dfrac{1}{4}x + \dfrac{1}{16} \displaystyle\int_0^x (x - t)u(t)dt$

5. $u(x) = 2 - \displaystyle\int_0^x (x - t)u(t)dt$

6. $u(x) = 1 - \int_0^x 2tu(t)dt$

7. $u(x) = x + \int_0^x (x - t)u(t)dt$

8. $u(x) = 1 - \int_0^x (x - t)u(t)dt$

9. $u(x) = 1 + x - \int_0^x (x - t)u(t)dt$

10. $u(x) = 1 - x - \int_0^x (x - t)u(t)dt$

11. $u(x) = 2 - x + \int_0^x u(t)dt$

12. $u(x) = 1 - x - \frac{1}{2}x^2 + \int_0^x (x - t)u(t)dt$

3.7 The Method of Successive Substitutions

The method of successive substitutions, that will be used here, is completely identical to that used in Chapter 2 for handling Fredholm integral equations. In this method, we set $x = t$ and $t = t_1$ in the Volterra integral equation

$$u(x) = f(x) + \lambda \int_0^x K(x,t)u(t)dt, \tag{164}$$

to obtain

$$u(t) = f(t) + \lambda \int_0^t K(t,t_1)u(t_1)dt_1. \tag{165}$$

Replacing $u(t)$ at the right hand side of (164) by its obtained value given by (165) yields

$$u(x) = f(x) + \lambda \int_0^x K(x,t)f(t)dt$$
$$+ \lambda^2 \int_0^x K(x,t) \int_0^t K(t,t_1)u(t_1)dt_1 dt. \tag{166}$$

Substituting $x = t_1$ and $t = t_2$ in (164) we obtain

$$u(t_1) = f(t_1) + \lambda \int_0^{t_1} K(t_1,t_2)u(t_2)dt_2. \tag{167}$$

Substituting the value of $u(t_1)$ obtained in (167) into the right hand side of (166) leads to

$$
\begin{aligned}
u(x) \; = \; f(x) \; &+ \lambda \int_0^x K(x,t) f(t) dt \\
&+ \lambda^2 \int_0^x \int_0^t K(x,t) K(t,t_1) f(t_1) dt_1 dt \\
&+ \lambda^3 \int_0^x \int_0^t \int_0^{t_1} K(x,t) K(t,t_1) K(t_1,t_2) u(t_2) dt_2 dt_1 dt.
\end{aligned}
$$
(168)

Accordingly, the general series form for $u(x)$ can be rewritten as

$$
\begin{aligned}
u(x) \; = \; f(x) &+ \lambda \int_0^x K(x,t) f(t) dt \\
&+ \lambda^2 \int_0^x \int_0^t K(x,t) K(t,t_1) f(t_1) dt_1 dt \\
&+ \lambda^3 \int_0^x \int_0^t \int_0^{t_1} K(x,t) K(t,t_1) K(t_1,t_2) f(t_2) dt_2 dt_1 dt \\
&+ \cdots
\end{aligned}
$$
(169)

We remark here that in this method the unknown function $u(x)$ is substituted by the given function $f(x)$ that makes the evaluation of the multiple integrals easily computable. This substitution of $u(x)$ occurs several times through the integrals and hence this is why it is called the method of successive substitutions. The technique will be illustrated by discussing the following examples.

Example 1. We solve the following Volterra integral equation

$$
u(x) = x - \int_0^x (x - t) u(t) dt,
$$
(170)

by using the method of successive substitutions. Substituting $\lambda = -1$, $f(x) = x$, and $K(x,t) = (x - t)$ into (169) yields

$$
u(x) = x - \int_0^x (x - t) t \, dt + \int_0^x \int_0^t (x - t)(t - t_1) t_1 \, dt_1 dt + \cdots,
$$
(171)

or equivalently

$$
u(x) = x - \int_0^x (xt - t^2) dt + \int_0^x \int_0^t (x - t)(t t_1 - t_1^2) dt_1 dt + \cdots.
$$
(172)

Therefore we obtain the solution in a series form

$$
u(x) = x - \frac{1}{3!} x^3 + \frac{1}{5!} x^5 + \cdots,
$$
(173)

or in a closed form

$$u(x) = \sin x \tag{174}$$

upon using the Taylor expansion for $\sin x$.

Example 2. We next solve the Volterra integral equation by applying the method of successive substitutions

$$u(x) = x^2 - x^4 + \int_0^x 4t\, u(t)dt. \tag{175}$$

Substituting $\lambda = 4$, $f(x) = x^2 - x^4$, and $K(x,t) = t$ into (169) yields

$$
\begin{aligned}
u(x) &= x^2 - x^4 + 4\int_0^x t\left(t^2 - t^4\right)dt + 16\int_0^x \int_0^t tt_1\left(t_1^2 - t_1^4\right)dt_1 dt \\
&\quad + 64\int_0^x \int_0^t \int_0^{t_1} tt_1 t_2\left(t_2^2 - t_2^4\right)dt_2 dt_1 dt \\
&\quad + \cdots,
\end{aligned}
\tag{176}
$$

and this will yield

$$u(x) = x^2 - x^4 + x^4 - \frac{2}{3}x^6 + \frac{2}{3}x^6 - \frac{1}{3}x^8 + \frac{1}{3}x^8 + \cdots. \tag{177}$$

Consequently, we easily obtain the exact solution

$$u(x) = x^2, \tag{178}$$

upon cancelling the similar terms with opposite signs in (177).

Exercises 3.7

Solve the following Volterra integral equations by the *successive substitutions method:*

1. $u(x) = x + \displaystyle\int_0^x u(t)dt$

2. $u(x) = \dfrac{1}{2!}x^2 + \displaystyle\int_0^x u(t)dt$

3. $u(x) = \dfrac{1}{3!}x^3 - \displaystyle\int_0^x (x-t)u(t)dt$

4. $u(x) = \dfrac{1}{3!}x^3 + \displaystyle\int_0^x (x-t)u(t)dt$

5. $u(x) = \dfrac{1}{2!}x^2 - \displaystyle\int_0^x (x-t)u(t)dt$

6. $u(x) = 1 - \dfrac{1}{2!}x^2 - \displaystyle\int_0^x u(t)dt$

7. $u(x) = 1 + 2 \int_0^x u(t)dt$

8. $u(x) = 3 - 2x + \int_0^x u(t)dt$

9. $u(x) = 2 + \frac{1}{2}x^2 - \int_0^x (x - t)u(t)dt$

10. $u(x) = 1 - x + \frac{1}{2}x^2 - \int_0^x (x - t)u(t)dt$

11. $u(x) = 2 - \frac{1}{2}x^2 + \int_0^x (x - t)u(t)dt$

12. $u(x) = \frac{1}{2!}x^2 + \int_0^x (x - t)u(t)dt$

3.8   Comparison between Alternative Methods

Before making a comparison between all methods discussed in this chapter, it is convenient to point out that there are other techniques for solving Volterra integral equations that are beyond the scope of this text. Using Laplace transforms to handle Volterra equations requires extensive background in Laplace transforms and the convolution integral. However, when it comes to selecting a preferable method among the methods that were introduced in the previous sections, we cannot recommend a specific method.

Even though the method of reducing Volterra integral equation to initial value problem is rarely used, but it is the only method that may give directly the exact solution in a closed form. This is easily seen if the resulting initial value problem has constant coefficients. We may obtain the solution in a series form, similar to the results obtained by other methods, when the coefficients of the resulting initial value problem are functions of the independent variable x. The latter case will not be introduced here. A useful example has been discussed by Example 1 in Section 3.2.

However, we found that if the kernel $K(x, t)$ of the integral equation is a degenerate one that consists of a polynomial of one or two terms, the series method and the decomposition method might be the best choices because it minimize the volume of calculations. The series solution obtained by using these methods might yield the exact solution in a closed form or we may obtain an approximation of the solution. Moreover, if $f(x)$ is a transcendental function, the series solution method works easier than the decomposition method.

Comparing the Adomian decomposition method with the successive approximation method, it is evident the decomposition method is much easier in that we integrate always very few terms to obtain the successive components, whereas in the other method we integrate many terms to evaluate the successive approximations after selecting the zeroth approximation. Moreover, for linear Volterra integral equations, the variational iteration method works effectively as the Adomian method.

In closing this section, we point out that the method of successive substitutions suffers from the huge size of calculations, especially if the function $f(x)$ is a trigonometric or exponential function.

To achieve our goal of the comparison between these methods, we illustrate this comparison by solving the following Volterra integral equation by using all various methods.

Example 1. We solve the following example

$$u(x) = 1 + \int_0^x u(t)dt, \tag{179}$$

by using the six alternative methods.

We will start with the Adomian decomposition method, then we use the variational method, the series solution method and so on for the other methods examined in this text.

(a) Adomian Decomposition Method:

As discussed earlier we set

$$u(x) = \sum_{n=0}^{\infty} u_n(x). \tag{180}$$

Substituting (180) into both sides of (179) we obtain

$$u_0(x) + u_1(x) + u_2(x) + \cdots = 1 + \int_0^x (u_0(t) + u_1(t) + u_2(t) + \cdots) \, dt. \tag{181}$$

As stated before, we have to set the zeroth component $u_0(x)$ by all terms outside the integral sign, hence we have

$$u_0(x) = 1. \tag{182}$$

The first component $u_1(x)$ may be obtained by

$$u_1(x) = \int_0^x u_0(t)dt, \tag{183}$$

which gives the first component

$$u_1(x) = x. \tag{184}$$

Proceeding in the same manner we can easily obtain

$$u_2(x) = \frac{1}{2!}x^2, \tag{185}$$

and so on for other components. Noting that in the decomposition method we have

$$u(x) = u_0 + u_1 + u_2 + u_3 + \cdots, \tag{186}$$

hence by using the results (182)-(185) into (186) we obtain the solution in a series form

$$u(x) = 1 + x + \frac{1}{2!}x^2 + \frac{1}{3!}x^3 + \cdots, \tag{187}$$

and in a closed form by the exact solution

$$u(x) = e^x. \tag{188}$$

(b) The Variational Iteration Method

Differentiating both sides of the integral equation (179) with respect to x gives

$$u'(x) = u(x). \tag{189}$$

Using the correction functional

$$u_{n+1}(x) = u_n(x) - \int_0^x \left(u_n'(\xi) - u_n(\xi) \right) d\xi, n \geq 0, \tag{190}$$

gives the following approximations

$$
\begin{aligned}
u_0(x) &= 1, \\
u_1(x) &= 1 - \int_0^x \left(u_0'(t) - u_0(t) \right) dt \\
&= 1 + x, \\
u_2(x) &= 1 + x - \int_0^x \left(u_1'(t) - u_1(t) \right) dt \\
&= 1 + x + \frac{1}{2!}x^2 \\
u_3(x) &= 1 + x + \frac{1}{2!}x^2 + \frac{1}{3!}x^3,
\end{aligned} \tag{191}
$$

and so on. The solution in a series form is thus given by

$$u(x) = 1 + x + \frac{1}{2!}x^2 + \frac{1}{3!}x^3 + \cdots \tag{192}$$

that converges to the exact solution

$$u(x) = e^x. \tag{193}$$

(c) The Series Method:

As indicated before, assuming that $u(x)$ is analytic, hence we may write

$$u(x) = \sum_{n=0}^{\infty} a_n x^n. \tag{194}$$

Substituting (194) into both sides of (179) we find

$$a_0 + a_1 x + a_2 x^2 + a_3 x^3 + \cdots = 1 + \int_0^x \left(a_0 + a_1 t + a_2 t^2 + \cdots \right) dt. \tag{195}$$

Integrating the easy integrals in the right hand side we obtain

$$a_0 + a_1 x + a_2 x^2 + a_3 x^3 + \cdots = 1 + a_0 x + \frac{1}{2} a_1 x^2 + \frac{1}{3} a_2 x^3 + \cdots. \tag{196}$$

Equating the coefficients of like powers of x from both sides we find

$$a_0 = 1, \tag{197}$$
$$a_1 = 1, \tag{198}$$
$$a_2 = \frac{1}{2!}, \tag{199}$$
$$a_3 = \frac{1}{3!}, \tag{200}$$

and so on. Substituting the results obtained for a_k, $k \geq 0$ into (194) we obtain the solution in a series form

$$u(x) = \sum_{n=0}^{\infty} \frac{x^n}{n!}, \tag{201}$$

and in a closed form

$$u(x) = e^x, \tag{202}$$

is the exact solution of the example under discussion.

(d) Converting to Initial Value Problems:

Differentiating both sides of (179) with respect to x we obtain

$$u'(x) = u(x), \tag{203}$$

with the initial condition

$$u(0) = 1. \tag{204}$$

Solving the first separable order differential equation (203) and using the initial condition (204) the exact solution

$$u(x) = e^x, \tag{205}$$

is readily obtained.

(e) Successive Approximations Method:

In this method we select the zeroth approximation by

$$u_0(x) = 1. \tag{206}$$

Following the technique that was discussed above, the other approximations of the solution $u(x)$ can be easily obtained by

$$u_1(x) \quad = \quad 1 + x, \tag{207}$$

$$u_2(x) \quad = \quad 1 + x + \frac{x^2}{2!}, \tag{208}$$

$$u_3(x) \quad = \quad 1 + x + \frac{x^2}{2!} + \frac{x^3}{3!}, \tag{209}$$

and so on. Accordingly, the nth component is given by

$$u_n(x) = 1 + x + \frac{x^2}{2!} + \frac{x^3}{3!} + \cdots + \frac{x^n}{n!}. \tag{210}$$

Consequently we find

$$u(x) \quad = \quad \lim_{n \to \infty} u_n(x),$$

$$= \quad \lim_{n \to \infty} 1 + x + \frac{x^2}{2!} + \frac{x^3}{3!} + \cdots \tag{211}$$

$$= \quad e^x,$$

the same solution obtained above.

(f) The Method of Successive Substitutions:

In this method we have to set $K(x,t) = 1$, $\lambda = 1$ and $f(x) = 1$, hence we have

$$
\begin{aligned}
u(x) &= 1 + 1 \int_0^x 1 dt + \int_0^x \int_0^t 1 dt_1 dt + \int_0^x \int_0^t \int_0^{t_1} 1 dt_2 dt_1 dt + \cdots \\
&= 1 + x + \frac{x^2}{2!} + \frac{x^3}{3!} + \cdots \\
&= e^x,
\end{aligned}
$$

(212)

the same result obtained by other methods.

An important observation, and not a recommendation, can be made from the comparison performed above which suggests that the decomposition method, the variational iteration method, and the series method introduce promising improvements over other existing techniques.

We point out that there are other powerful techniques that are used for solving Volterra integral equations. Examples of these techniques include wavelet method, Legendre wavelets method, Laplace transform method, and many others.

3.9 Volterra Integral Equations of the First Kind

In this section we will study the Volterra integral equation of the first kind with separable kernel given by

$$
f(x) = \int_0^x K(x,t) \, u(t) dt.
$$

(213)

We point out that the Adomian decomposition method or the variational iteration method cannot be used in a straightforward manner. However, the traditional series solution method that we used in this chapter can be used in a direct manner. Although there are other analytical and numerical methods that can be used to handle Volterra integral equations of the first kind, but in this section we will concern ourselves with two practical methods.

3.9.1 The Series Solution Method

The method was presented earlier. In this method we use the Taylor series for the solution $u(x)$ in the form

$$
u(x) = \sum_{n=0}^{\infty} a_n x^n,
$$

(214)

where $a_n, n \geq 0$ are coefficients that will be determined. We then substitute this series into both sides of the Volterra integral equations of the first kind. We next evaluate the integral at the right side. By equating the coefficients of like terms into both sides we determine the coefficients $a_n, n \geq 0$, and this in turn leads to the solution in a series form. This will be illustrated by studying the following two examples.

Example 1. Find the solution of the Volterra equation of the first kind

$$x^2 + \frac{1}{6}x^3 = \int_0^x (2 + x - t)\, u(t)dt. \tag{215}$$

Substituting the series (214) in the right side and evaluating the resulting integral we find

$$x^2 + \frac{1}{6}x^3 = 2a_0 x + (\frac{1}{2}a_0 + a_1)x^2 + (\frac{1}{3!}a_1 + \frac{2}{3}a_2)x^3 + \cdots. \tag{216}$$

Equating the coefficients of like terms from both sides gives

$$\begin{aligned}
a_0 &= 0, \\
a_1 &= 1, \\
a_2 &= a_3 = a_4 = \cdots = 0.
\end{aligned} \tag{217}$$

Substituting this result into (214) gives the exact solution

$$u(x) = x. \tag{218}$$

Example 2. Find the solution of the Volterra equation of the first kind

$$xe^x = \int_0^x e^{x-t}\, u(t)dt, \tag{219}$$

or equivalently

$$x = \int_0^x e^{-t}\, u(t)dt. \tag{220}$$

Substituting the series (214) in the right side and evaluating the resulting integral we find

$$x = a_0 x + (-\frac{1}{2}a_0 + \frac{1}{2}a_1)x^2 + (\frac{1}{6}a_0 - \frac{1}{3}a_1 + \frac{1}{3}a_2)x^3 + \cdots. \tag{221}$$

Equating the coefficients of like terms from both sides gives

$$\begin{aligned}
a_0 &= 1, \\
a_1 &= 1, \\
a_2 &= \tfrac{1}{2!}, \\
a_3 &= \tfrac{1}{3!}, \\
&\ \vdots
\end{aligned} \tag{222}$$

Substituting this result into (214) gives the series solution

$$u(x) = 1 + x + \frac{1}{2!}x^2 + \frac{1}{3!}x^3 + \cdots, \tag{223}$$

that converges to the exact solution

$$u(x) = e^x. \tag{224}$$

To apply the Adomian method or the variational iteration method, we should convert the Volterra integral equation of the first kind to a Volterra integral equation of the second kind.

3.9.2 Conversion of First Kind to Second Kind

It is important to note that the Volterra integral equation of the first kind can be handled simply by converting this equation to a Volterra equation of the second kind. This goal can be accomplished by differentiating both sides of (213) with respect to x to obtain

$$f'(x) = K(x,x)u(x) + \int_0^x K_x(x,t)u(t)dt, \tag{225}$$

by using Leibniz rule. If $K(x,x) \neq 0$ in the interval of discussion, then dividing both sides of (225) by $K(x,x)$ yields

$$u(x) = \frac{f'(x)}{K(x,x)} - \frac{1}{K(x,x)} \int_0^x K_x(x,t)\, u(t)dt, \tag{226}$$

a Volterra integral equation of the second kind. The case where the kernel $K(x,x) = 0$ leads to a complicated behavior of the problem that will not be investigated here.

To solve (226) we select any method that we discussed before. The technique of differentiating both sides of Volterra integral equation of the first kind, verifying that $K(x,x) \neq 0$, reducing to Volterra integral equation of the second kind and solving the resulting equation will be illustrated by discussing the same two examples solved by using the series solution method.

Example 3. Find the solution of the Volterra equation of the first kind

$$x^2 + \frac{1}{6}x^3 = \int_0^x (2 + x - t)\, u(t)dt. \tag{227}$$

We note first that $K(x,t) = 2 + x - t$, hence $K(x,x) = 2 \neq 0$. Differentiating both sides of (227) with respect to x yields

$$2x + \frac{1}{2}x^2 = 2u(x) + \int_0^x u(t)dt, \tag{228}$$

or equivalently

$$u(x) = x + \frac{1}{4}x^2 - \frac{1}{2}\int_0^x u(t)dt. \tag{229}$$

We prefer to use the *modified decomposition method*, hence we set

$$u_0(x) = x, \tag{230}$$

which gives

$$\begin{aligned}u_1(x) &= \frac{1}{4}x^2 - \frac{1}{2}\int_0^x t\,dt \\ &= 0.\end{aligned} \tag{231}$$

Accordingly, other components $u_n(x) = 0$, $n \geq 2$. The exact solution $u(x)$ is given by

$$u(x) = x, \tag{232}$$

obtained upon using the components obtained above.

Example 4. Find the solution of the Volterra equation of the first kind

$$xe^x = \int_0^x e^{x-t} u(t)dt. \tag{233}$$

We note first that $K(x,t) = e^{x-t}$, hence $K(x,x) = 1 \neq 0$. Differentiating both sides of (233) with respect to x yields

$$e^x + xe^x = u(x) + \int_0^x e^{x-t}u(t)dt, \tag{234}$$

or equivalently

$$u(x) = e^x + xe^x - \int_0^x e^{x-t}u(t)\,dt. \tag{235}$$

We shall solve the resulting equation by the *Adomian decomposition method* and by the *modified decomposition method*. We first start by applying the Adomian decomposition method, therefore we set

$$u_0(x) = e^x + xe^x. \tag{236}$$

Consequently, we obtain

$$\begin{aligned}u_1(x) &= -\int_0^x e^{x-t}\left(e^t + te^t\right)dt \\ &= -e^x\left(x + \frac{x^2}{2!}\right),\end{aligned} \tag{237}$$

and

$$u_2(x) = \int_0^x e^{x-t}\left(te^t + \frac{t^2}{2!}e^t\right)dt$$
$$= e^x\left(\frac{x^2}{2!} + \frac{x^3}{3!}\right),$$

(238)

and so on. Using the above results of the components obtained gives

$$u(x) = e^x\left(1 + x - x - \frac{x^2}{2!} + \frac{x^2}{2!} + \frac{x^3}{3!} - \frac{x^3}{3!} - \cdots\right)dt$$
$$= e^x,$$

(239)

the exact solution obtained upon cancelling like terms with opposite signs.
Using the *modified decomposition method* we set

$$u_0(x) = e^x,$$

(240)

which gives

$$u_1(x) = xe^x - \int_0^x e^x\,dt,$$
$$= 0.$$

(241)

It immediately follows that

$$u_n(x) = 0, \ n \geq 1.$$

(242)

Accordingly, the exact solution is

$$u(x) = e^x.$$

(243)

Exercises 3.9

Solve the following Volterra integral equations of the first kind by any method:

1. $5x^2 + x^3 = \int_0^x (5 + 3x - 3t)u(t)dt.$

2. $xe^{-x} = \int_0^x e^{t-x}u(t)dt.$

3. $2e^x - x - 2 = \int_0^x (1 + x - t)u(t)dt.$

4. $2\cosh x - \sinh x - (2 - x) = \int_0^x (2 - x + t)u(t)dt.$

5. $4\sin x - 3\cos x + 3 = \int_0^x (4 + 3x - 3t)u(t)dt.$

6. $\tan x - \ln(\cos x) = \int_0^x (1 + x - t)u(t)dt, \quad x < \pi/2.$

Chapter 4

Fredholm Integro-Differential Equations

4.1 Introduction

In this chapter we will be concerned with the Fredholm integro-differential equations where both differential and integral operators will appear in the same equation. The differential operator may be of first order or higher order.

Scientists and researchers investigated the topic of integro-differential equations through their work in science applications such as heat transfer, diffusion process in general, neutron diffusion and biological species coexisting together with increasing and decreasing rates of generating. More details about the sources where these equations arise can be found in physics, biology and engineering applications as well as in advanced integral equations books.

In the Fredholm integro-differential equations, it is important to note that the unknown function $u(x)$ and one or more of its derivatives such as $u'(x)$, $u''(x)$, ..., appear outside and inside the integral sign as well. The Fredholm integro-differential equations come as a first kind and as a second kind as defined for the Fredholm integral equations.

The following are examples of linear integro-differential equations:

$$
\begin{aligned}
u'(x) &= x - \int_0^1 e^{x-t}u(t)dt, \quad u(0) = 0, \\
u''(x) &= e^x - x + \int_0^1 xtu'(t)dt, \quad u(0) = 1, u'(0) = 1.
\end{aligned}
\tag{1}
$$

It is clear from the examples given above that the unknown function $u(x)$ or one of its derivatives appear under the integral sign, and other derivatives of $u(x)$ appear out of the integral sign as well. Therefore, the above given equations involve the derivatives and the integral operators in the same equation, and consequently the term integro-differential equations has been used for problems involving this combination of operators. Moreover, because these equations involve differential operator, it is necessary to describe the initial conditions for each problem,

To determine a solution for the Fredholm integro-differential equation, the initial conditions should be given as stated earlier, and this may be clearly seen as a result of involving $u(x)$ and its derivatives. The initial conditions in solving any differential equation are needed to determine the constants of integration.

In what follows, we will discuss several methods that can handle successfully the linear Fredholm integro-differential equations of the second kind where the solution $u(x)$ and at least one of its derivatives appear inside and outside the integral sign.

4.2   Fredholm Integro-Differential Equations

In this section we will discuss the reliable methods used to solve Fredholm integro-differential equations. This type of equations was termed as Fredholm integro-differential equations, given in the form

$$
u^{(n)}(x) = f(x) + \lambda \int_a^b K(x,t)u(t)dt.
\tag{2}
$$

We remark here that we will focus our concern on the equations that involve separable kernels where the kernel $K(x,t)$ can be expressed as a finite sum of the form

$$
K(x,t) = \sum_{k=1}^{n} g_k(x) h_k(t).
\tag{3}
$$

Without loss of generality, we will make our analysis on a one term kernel $K(x,t)$ of the form

$$
K(x,t) = g(x) h(t),
\tag{4}
$$

and this can be generalized for other cases. The non-separable kernel can be reduced to separable kernel by using the Taylor expansion for the kernel involved. The Fredholm integro-differential equations are usually solved by a variety of methods, some are numerical methods, whereas others are analytic methods. It is worth noting that in this chapter we will introduce the most recent and practical schemes that handle this type of equations, where we may obtain an exact solution or an approximation to the solution with the highest desirable accuracy. We point out here that the methods to be discussed are introduced before, but we will focus our discussion on how these methods can be implemented in this type of equations. We first start with the most practical traditional method.

We point out that the direct computation method, that will be used first, requires integrating or multiple integrating of the derivatives that are involved in the equation. In what follows, we list some of the facts used in calculus courses:

$$
\begin{array}{rcl}
\int_0^x u'(t)\, dt &=& u(x) - u(0), \\
\int_0^x \int_0^t u''(s)\, ds\, dt &=& u'(x) - u(0) - xu'(0), \\
\int_0^x \int_0^t \int_0^s u'''(r)\, dr\, ds\, dt &=& u'(x) - u(0) - xu'(0) - \frac{1}{2}x^2 u''(0),
\end{array}
\tag{5}
$$

and so on for other derivatives.

4.3 The Direct Computation Method

The direct computation method has been extensively used in Chapter 2 to handle Fredholm integral equations. Without loss of generality, we may assume a standard form to the Fredholm integro-differential equation given by

$$
u^{(n)}(x) = f(x) + \int_0^1 K(x,t)\, u(t)dt, \quad u^{(k)}(0) = b_k,\ 0 \le k \le (n-1) \tag{6}
$$

where $u^{(n)}(x)$ indicates the nth derivative of $u(x)$ with respect to x and b_k are constants that define the initial conditions. Substituting (4) into (6) yields

$$
u^{(n)}(x) = f(x) + g(x) \int_0^1 h(t)\, u(t)dt, \quad u^{(k)}(0) = b_k,\ 0 \le k \le (n-1). \tag{7}
$$

We can easily observe that the definite integral in the integro-differential equation (7) involves an integrand that completely depends on the variable t, and therefore, it seems reasonable to set that definite integral in the right side of (7) to a constant α, i.e. we set

$$\alpha = \int_0^1 h(t)u(t)\,dt. \tag{8}$$

With α defined in (8), the equation (7) can be written by

$$u^{(n)}(x) = f(x) + \alpha\,g(x). \tag{9}$$

It remains to determine the constant α to evaluate the exact solution $u(x)$. To find α, we should derive a form for $u(x)$ by using (9), followed by substituting this form in (8). To achieve this we integrate both sides of (9) n times from 0 to x, and by using the given initial conditions $u^{(k)}(0) = b_k$, $0 \le k \le (n-1)$ we obtain an expression for $u(x)$ given by

$$u(x) = p(x;\alpha), \tag{10}$$

where $p(x;\alpha)$ is the result derived from integrating (9) and by using the given initial conditions. Substituting (10) into the right hand side of (8), integrating and solving the resulting equation lead to a complete determination of α. The exact solution of (6) follows immediately upon substituting the resulting value of α into (10).

To give a clear view of the technique, we illustrate the method by solving the following examples.

Example 1. Solve the first-order Fredholm integro-differential equation by using the direct computation method:

$$u'(x) = 1 - \frac{1}{3}x + x\int_0^1 t\,u(t)\,dt, \quad u(0) = 0. \tag{11}$$

The equation (11) may be written in the form

$$u'(x) = 1 - \frac{1}{3}x + \alpha\,x, \quad u(0) = 0, \tag{12}$$

where the constant α is defined by

$$\alpha = \int_0^1 t\,u(t)\,dt. \tag{13}$$

To determine α, we first need an expression for $u(x)$ to be used in (13). This can be easily done by integrating both sides of (12) from 0 to x, using (5), and by using the given initial condition we obtain

$$u(x) = x + \left(\frac{\alpha}{2} - \frac{1}{6}\right)x^2. \tag{14}$$

Substituting (14) into (13) and evaluating the integral we find

$$\alpha = \frac{1}{3}, \tag{15}$$

so that the exact solution

$$u(x) = x, \tag{16}$$

follows immediately upon using (15) into (14).

Example 2. Solve the following third-order Fredholm integro-differential equation by using the direct computation method:

$$u'''(x) = \sin x - x - \int_0^{\pi/2} xt\, u'(t)dt, \tag{17}$$

subject to the initial conditions

$$u(0) = 1, \quad u'(0) = 0, \quad u''(0) = -1. \tag{18}$$

This equation can be written in the form

$$u'''(x) = \sin x - (1+\alpha)x, \quad u(0) = 1, \quad u'(0) = 0, \quad u''(0) = -1, \tag{19}$$

where

$$\alpha = \int_0^{\pi/2} t\, u'(t)dt. \tag{20}$$

To determine α, we should find an expression for $u'(x)$ in terms of x and α to be used in (20). This can be achieved by integrating (19) three times from 0 to x, using (5), and using the initial conditions, hence we find

$$u''(x) = -\cos x - \frac{1+\alpha}{2!}x^2, \tag{21}$$

$$u'(x) = -\sin x - \frac{1+\alpha}{3!}x^3, \tag{22}$$

and

$$u(x) = \cos x - \frac{1+\alpha}{4!}x^4. \tag{23}$$

Substituting (22) into (20) we obtain

$$\alpha = \int_0^{\pi/2} \left(-t\sin t - \frac{1+\alpha}{3!}t^4 \right) dt, \tag{24}$$

which gives

$$\alpha = -1. \tag{25}$$

Substituting (25) into (23) gives the exact solution

$$u(x) = \cos x. \tag{26}$$

Example 3. Solve the following second-order Fredholm integro-differential equation by using the direct computation method:

$$u''(x) = 2 - \frac{2}{3}x - \frac{16}{15}x^2 + \int_{-1}^{1} (xt^2 + x^2t^2)\, u(t)dt, \tag{27}$$

subject to the initial conditions

$$u(0) = 1, \quad u'(0) = 1. \tag{28}$$

This equation can be written in the form

$$u''(x) = 2 + (\alpha - \frac{2}{3}x) + (\beta - \frac{16}{15})x^2, \quad u(0) = 1, \quad u'(0) = 1 \tag{29}$$

where

$$\alpha = \int_{-1}^{1} tu(t)\, dt,$$
$$\beta = \int_{-1}^{1} t^2u(t)\, dt. \tag{30}$$

To determine α, and β, we should find an expression for $u(x)$ in terms of x, α and β to be used in (30). This can be achieved by integrating (29) two times from 0 to x, using (5), and using the initial conditions, hence we find

$$u(x) = 1 + x + x^2 + (\alpha - \frac{2}{3})\frac{x^3}{6} + (\beta - \frac{16}{15})\frac{x^4}{12}. \tag{31}$$

Substituting (31) into (30), solving for α and β, we obtain

$$\alpha = \frac{2}{3}, \; \beta = \frac{16}{15}, \tag{32}$$

which gives the exact solution

$$u(x) = 1 + x + x^2. \tag{33}$$

Exercises 4.3

Solve the following Fredholm integro-differential equations by using the *direct computation method*

1. $u'(x) = \frac{1}{6} + \frac{5}{36}x - \int_{0}^{1} xtu(t)\, dt, \quad u(0) = \frac{1}{6}.$

2. $u'(x) = \frac{1}{21}x - \int_{0}^{1} xtu(t)\, dt, \quad u(0) = \frac{1}{6}.$

3. $u''(x) = -\sin x + x - \int_0^{\pi/2} xtu(t)\, dt, \quad u(0) = 0, u'(0) = 1.$

4. $u''(x) = \dfrac{9}{4} - \dfrac{1}{3}x + \int_0^1 (x-t)u(t)\, dt, \quad u(0) = u'(0) = 0.$

5. $u'(x) = 2\sec x^2 \tan x - x + \int_0^{\pi/4} xu(t)\, dt, \quad u(0) = 1.$

6. $u'(x) = -5 - 6x + \int_{-1}^1 (x-t)u(t)\, dt, \quad u(0) = 1.$

7. $u'(x) = \cos x + 1 - x + \int_0^{\pi/2} (x-t)u(t)\, dt, \quad u(0) = 0.$

8. $u'(x) = \cos x - \sin x - 2x + \dfrac{\pi}{2} + \int_0^{\pi/2} (x-t)u(t)\, dt, \quad u(0) = 1.$

4.4 The Adomian Decomposition Method

The Adomian decomposition method in its simplest form, has been extensively introduced in Chapters 2 and 3 for handling Fredholm and Volterra integral equations. In this section we will study how this powerful method can be implemented to determine a series solution to the Fredholm integro-differential equations. As indicated earlier, we may assume a standard form to the Fredholm integro-differential equation given by

$$u^{(n)}(x) = f(x) + \int_0^1 K(x,t)\, u(t)dt, \quad u^{(k)}(0) = b_k, \ 0 \le k \le (n-1) \quad (34)$$

where $u^{(n)}(x)$ indicates the nth derivative of $u(x)$ with respect to x and b_k are constants that give the initial conditions. Substituting (4) into (34) yields

$$u^{(n)}(x) = f(x) + g(x) \int_0^1 h(t)\, u(t)dt. \quad (35)$$

We can easily observe that the definite integral in the integro-differential equation (35) involves an integrand that completely depends on the variable t as discussed in the preceding section. In an operator form, the equation (35) can be written as

$$Lu(x) = f(x) + g(x) \int_0^1 h(t)\, u(t)dt, \quad (36)$$

where the differential operator L is given by

$$L = \frac{d^n}{dx^n}. \quad (37)$$

It is clear that L is an invertible operator, therefore the integral operator L^{-1} is an *n-fold* integration operator and may be considered as definite integrals from 0 to x for each integral. Applying L^{-1} to both sides of (36) yields

$$
\begin{aligned}
u(x) = b_0 + b_1 x + \frac{1}{2!} b_2 x^2 + \cdots + \frac{1}{(n-1)!} b_{n-1} x^{n-1} + L^{-1}\left(f(x)\right) \\
+ \left(\int_0^1 h(t)\, u(t) dt \right) L^{-1}\left(g(x)\right).
\end{aligned}
\tag{38}
$$

In other words we integrated (35) n times from 0 to x and we used the initial conditions at every step of integration. It is important to note that the equation obtained in (38) is a standard Fredholm integral equation. This note will be used in the coming section.

In the decomposition method we usually define the solution $u(x)$ of (34) in a series form given by

$$
u(x) = \sum_{n=0}^{\infty} u_n(x).
\tag{39}
$$

Substituting (39) into both sides of (38) we get

$$
\begin{aligned}
\sum_{n=0}^{\infty} u_n(x) \;=\; & \sum_{k=0}^{n-1} \frac{1}{k!} b_k x^k + L^{-1}\left(f(x)\right) \\
& + \left(\int_0^1 h(t)\, (\sum_{n=0}^{\infty} u_n(t))\, dt \right) L^{-1}\left(g(x)\right)
\end{aligned}
\tag{40}
$$

or equivalently

$$
\begin{aligned}
u_0(x) + u_1(x) + u_2(x) + \cdots \;=\; & \sum_{k=0}^{n-1} \frac{1}{k!} b_k x^k + L^{-1}\left(f(x)\right) \\
& + \left(\int_0^1 h(t)\, u_0(t) dt \right) L^{-1}\left(g(x)\right) \\
& + \left(\int_0^1 h(t)\, u_1(t) dt \right) L^{-1}\left(g(x)\right) \\
& + \left(\int_0^1 h(t)\, u_2(t) dt \right) L^{-1}\left(g(x)\right) \\
& + \cdots
\end{aligned}
\tag{41}
$$

The components $u_i(x), i \geq 0$ of the unknown function $u(x)$ are determined in a recurrent manner, in a similar fashion as discussed before, where we set

$$u_0(x) = \sum_{k=0}^{n-1} \frac{1}{k!} b_k x^k + L^{-1}\left(f(x)\right) \tag{42}$$

$$u_1(x) = \left(\int_0^1 h(t)\, u_0(t)\, dt\right) L^{-1}\left(g(x)\right) \tag{43}$$

$$u_2(x) = \left(\int_0^1 h(t)\, u_1(t)\, dt\right) L^{-1}\left(g(x)\right) \tag{44}$$

$$u_3(x) = \left(\int_0^1 h(t)\, u_2(t)\, dt\right) L^{-1}\left(g(x)\right) \tag{45}$$

and so on. The above discussed scheme for the determination of the components $u_0(x), u_1(x), u_2(x), u_3(x), \ldots$ of the solution $u(x)$ of the equation (34) can be written in a recursive relationship by

$$u_0(x) = \sum_{k=0}^{n-1} \frac{1}{k!} b_k x^k + L^{-1}\left(f(x)\right)$$

$$\tag{46}$$

$$u_{n+1}(x) = \left(\int_0^1 h(t)\, u_n(t)\, dt\right) L^{-1}\left(g(x)\right), \quad n \geq 0.$$

In view of (46), the components $u_0(x)$, $u_1(x)$, $u_2(x)$, $u_3(x)$, $u_4(x)$, ... of $u(x)$ are immediately determined. With these components established, the solution $u(x)$ of (34) is readily determined in a series form using (39). Consequently, the series obtained for $u(x)$ frequently provides the exact solution in a closed form as will be illustrated later. However, for some problems, where a closed form is not easy to find, we use the series form obtained to approximate the solution. We point out here that few terms of the series derived by the decomposition method usually provide the higher accuracy level of the approximate solution.

The decomposition method avoids massive computational work and difficulties that arise from other methods. The computational work can be minimized, sometimes, by using the modified decomposition method or by observing the so-called self-cancelling noise terms phenomena. The aforementioned techniques were presented earlier in details, hence it will be summarized in the following sections.

4.4.1   The Modified Decomposition Method

The modified decomposition method is a powerful technique that minimizes the size of calculations. An essential requirement for the use of this method is that the data function $f(x)$ in (36) should consist of more than one term. Consequently, the data function $f(x)$ can be decomposed to two parts, as introduced in [59], of the form

$$f(x) = f_0(x) + f_1(x). \tag{47}$$

As stated before, the selection of these two parts depends mainly on a trial basis. Moreover, using this selection, we set the modified recurrence relation as

$$
\begin{aligned}
u_0(x) &= \sum_{k=0}^{n-1} \frac{1}{k!} b_k x^k + L^{-1}(f_0(x)) \\
u_1(x) &= L^{-1}(f_1(x)) + \left(\int_0^1 h(t)\, u_0(t) dt \right) L^{-1}(g(x)), \\
u_{n+1}(x) &= \left(\int_0^1 h(t)\, u_n(t) dt \right) L^{-1}(g(x)), \quad n \geq 1.
\end{aligned}
\tag{48}
$$

Although the change between the Adomian decomposition method is slight, but it was proved from using the proposed modification that this techniques minimizes the calculations size, and mostly exact solution can be derived in using two iterations only for $u_0(x)$ and $u_1(x)$. The strength of this modified method was confirmed before in Chapters 2 and 3.

Another observation that facilitates the convergence of the solution is the so-called noise terms phenomenon that was explained before, but will be summarized next.

4.4.2   The Noise Terms Phenomenon

The noise terms phenomenon, that handles the self-cancelling noise terms, was introduced in previous chapters and used effectively in the literature. It was proved by Adomian and Rach [4] and others that the exact solution of any integral or integro-differential equation, for some cases, may be obtained by considering the first two components u_0 and u_1 only. Instead of evaluating several components, it is useful to examine the first two components u_0 and u_1. The conclusion made in [4] suggests that if we observe the appearance of like terms in both components with opposite signs, then by cancelling these terms, the remaining non-cancelled terms of u_0 may in some cases provide the exact solution. This can be justified through substitution. The self-cancelling terms, the identical terms with opposite terms, between the components u_0 and u_1 are called the noise terms. It was formally proved that other terms in other components will vanish in the limit if the noise terms occurred in $u_0(x)$ and $u_1(x)$. However, if the exact solution was not attainable by using this phenomena, then we should continue determining other components of $u(x)$ to get a closed form solution or an approximate solution as discussed earlier.

Moreover, it is important to note that, even though this is a remarkable achievement that speeds the convergence of the solution and minimizes the size of calculations work, but unfortunately the self cancelling noise terms do not appear always, but a necessary condition should hold for the possibility that these terms may appear. The condition of the appearance of the noise terms, as proved in [62], requires the existence of the exact solution as one term of the zeroth component $u_0(x)$.

In the following we discuss some examples which illustrate the above outlined decomposition scheme, the modified decomposition method, and the phenomenon of the self cancelling noise terms as well.

Example 1. Solve the following Fredholm integro-differential equation

$$u'(x) = \cos x + \frac{1}{4}x - \frac{1}{4}\int_0^{\pi/2} xt\, u(t)dt, \quad u(0) = 0. \qquad (49)$$

Integrating both sides of the equation (49) from 0 to x gives

$$u(x) - u(0) = \sin x + \frac{1}{8}x^2 - \frac{1}{8}x^2 \int_0^{\pi/2} t\, u(t)dt, \quad u(0) = 0, \qquad (50)$$

which gives upon using the initial condition

$$u(x) = \sin x + \frac{1}{8}x^2 - \frac{1}{8}x^2 \int_0^{\pi/2} t\, u(t)dt. \qquad (51)$$

(i) *Using the Adomian decomposition method:* We usually decompose the solution into a series form given by

$$u(x) = \sum_{n=0}^{\infty} u_n(x). \qquad (52)$$

Substituting (52) into both sides of (51) yields

$$\sum_{n=0}^{\infty} u_n(x) = \sin x + \frac{1}{8}x^2 - \frac{1}{8}x^2 \int_0^{\pi/2} t\left(\sum_{n=0}^{\infty} u_n(t)\right)dt. \qquad (53)$$

The Adomian decomposition method admits the use of the recurrence relation

$$u_0(x) = \sin x + \frac{1}{8}x^2,$$

$$u_1(x) = -\frac{1}{8}x^2 \left(\int_0^{\pi/2} t\, u_0(t)dt \right) = -\frac{1}{8}x^2 - \frac{\pi^4}{16^3}x^2,$$

$$u_2(x) = -\frac{1}{8}x^2 \left(\int_0^{\pi/2} t\, u_1(t)dt \right) = \frac{\pi^4}{16^3}x^2 + \frac{\pi^8}{2 \times 16^5}x^2 \qquad (54)$$

$$u_3(x) \quad -\frac{1}{8}x^2 \left(\int_0^{\pi/2} t\, u_2(t)dt \right) = -\frac{\pi^8}{2 \times 16^3}x^2 + \text{other terms}$$

$$+ \cdots.$$

In view of these results, we obtain the exact solution

$$u(x) = \sin x. \qquad (55)$$

(ii) *Using the noise terms phenomenon:* Using the first two components $u_0(x)$ and $u_1(x)$, we observe that the two identical terms $\frac{1}{8}x^2$ appear in these two components with opposite signs. Cancelling this noise term from $u_0(x)$, the remaining non-cancelled term in $u_0(x)$ gives the exact solution

$$u(x) = \sin x, \qquad (56)$$

that should be checked if it justifies the integro-differential equation.

(iii) *Using the modified decomposition method:* We first decompose the data function $f(x) = \sin x + \frac{1}{8}x^2$ as follows:

$$\begin{aligned} f_0(x) &= \sin x, \\ f_1(x) &= \frac{1}{8}x^2, \end{aligned} \qquad (57)$$

using the modified recurrence relation (48) gives

$$\begin{aligned} u_0(x) &= \sin x, \\ u_1(x) &= \frac{1}{8}x^2 - \frac{1}{8}x^2 \left(\int_0^{\pi/2} t\, u_0(t)dt \right) = 0, \end{aligned} \qquad (58)$$

that gives the exact solution (56) that we justified.

Example 2. Solve the following Fredholm integro-differential equation

$$u'(x) = \frac{1}{6} - \frac{1}{18}x + \int_0^1 xt\, u(t)dt, \quad u(0) = 0. \qquad (59)$$

(i) *Using the Adomian decomposition method:* Integrating both sides of the equation (59) from 0 to x gives

$$u(x) - u(0) = \frac{1}{6}x - \frac{1}{36}x^2 + \frac{1}{2}x^2 \left(\int_0^1 t\, u(t)dt \right), \quad u(0) = 0, \qquad (60)$$

which gives upon using the initial condition

$$u(x) = \frac{1}{6}x - \frac{1}{36}x^2 + \frac{1}{2}x^2 \left(\int_0^1 t\, u(t)dt \right). \qquad (61)$$

In the decomposition method we usually express the solution $u(x)$ into a series form given by

$$u(x) = \sum_{n=0}^{\infty} u_n(x). \qquad (62)$$

Substituting (62) into both sides of (61) gives

$$\sum_{n=0}^{\infty} u_n(x) = \frac{1}{6}x - \frac{1}{36}x^2 + \frac{1}{2}x^2 \left(\int_0^1 t \left(\sum_{n=0}^{\infty} u_n(t) \right) dt \right). \qquad (63)$$

The Adomian decomposition method admits the determination of the components in a recurrent manner as follows:

$$
\begin{aligned}
u_0(x) &= \frac{1}{6}x - \frac{1}{36}x^2, \\
u_1(x) &= \frac{1}{2}x^2 \left(\int_0^1 t\, u_0(t)dt \right) = \frac{7}{288}x^2, \\
u_2(x) &= \frac{1}{2}x^2 \left(\int_0^1 t\, u_1(t)dt \right) = \frac{7}{8 \times 288}x^2, \qquad (64) \\
u_3(x) &= \frac{1}{2}x^2 \left(\int_0^1 t\, u_2(t)dt \right) = \frac{7}{64 \times 288}x^2, \\
&\ \vdots
\end{aligned}
$$

Consequently the exact solution can be obtained upon using

$$u(x) = \frac{1}{6}x - \frac{1}{36}x^2 + \frac{7}{288}x^2 \left(1 + \frac{1}{8} + \frac{1}{64} + \cdots \right), \qquad (65)$$

which gives the exact solution

$$u(x) = \frac{1}{6}x, \qquad (66)$$

upon evaluating the sum of the infinite geometric series.

(ii) *Using the noise terms phenomenon:* From the preceding calculations
we find

$$
\begin{array}{rcl}
u_0(x) & = & \frac{1}{6}x - \frac{1}{36}x^2, \\
u_1(x) & = & \frac{7}{288}x^2.
\end{array}
\tag{67}
$$

Notice that $u_1(x)$ can be written as

$$
u_1(x) = \frac{1}{36}x^2 - \frac{1}{288}x^2.
\tag{68}
$$

The noise terms $\pm\frac{1}{36}x^2$ appear between the components $u_0(x)$ and $u_1(x)$.
By canceling the noise term from $u_0(x)$, we obtain the exact solution given
in (66).

(iii) *Using the modified decomposition method:* We first split $f(x)$ into two
parts, given by

$$
\begin{array}{rcl}
f_0(x) & = & \frac{1}{6}x, \\
f_1(x) & = & -\frac{1}{36}x^2.
\end{array}
\tag{69}
$$

We next use the modified recursion relation

$$
\begin{array}{rcl}
u_0(x) & = & \frac{1}{6}x - \frac{1}{36}x^2, \\
u_1(x) & = & -\frac{1}{36}x^2 + \frac{1}{x^2}\int_0^1 t u_0(t)\,dt = 0.
\end{array}
\tag{70}
$$

The other components vanish in the limit. This gives the exact solution
given earlier in (66).

Example 3. Solve the following Fredholm integro-differential equation

$$
u'''(x) = \sin x - x - \int_0^{\pi/2} xt\, u'(t)dt, \quad u(0) = 1,\ u'(0) = 0,\ u''(0) = -1.
\tag{71}
$$

Using the Adomian decomposition method: We point out here that the first
derivative $u'(x)$ of the unknown function $u(x)$ appears under the integral
sign in this example. The approach we will follow is the same as used
before, and will be illustrated through the solution. Integrating both sides
of the equation (71) three times from 0 to x yields

$$
u(x) - u(0) - \frac{1}{2!}x^2 u''(0) = \cos x + \frac{1}{2!}x^2 - \frac{1}{4!}x^4 - 1 - \frac{1}{4!}x^4\int_0^{\pi/2} t u'(t)dt,
\tag{72}
$$

which gives upon using the initial conditions in (71)

$$
u(x) = \cos x - \frac{1}{4!}x^4 - \frac{1}{4!}x^4\int_0^{\pi/2} t u'(t)dt.
\tag{73}
$$

We begin by expressing the solution $u(x)$ into a series form by

$$u(x) = \sum_{n=0}^{\infty} u_n(x). \tag{74}$$

Substituting (74) into both sides of (73) yields

$$\sum_{n=0}^{\infty} u_n(x) = \cos x - \frac{1}{4!}x^4 - \frac{1}{4!}x^4 \int_0^{\pi/2} t \left(\sum_{n=0}^{\infty} u_n'(t) \right) dt. \tag{75}$$

Proceeding as before, we determine the components $u_i(x), i \geq 0$ recurrently, where we set

$$
\begin{aligned}
u_0(x) &= \cos x - \frac{1}{4!}x^4. \\
u_1(x) &= -\frac{1}{4!}x^4 \left(\int_0^{\pi/2} t\, u_0'(t)dt \right) = \frac{1}{4!}x^4 + \frac{\pi^5}{(5!)(3!)(32)}x^4, \\
u_2(x) &= -\frac{1}{4!}x^4 \left(\int_0^{\pi/2} t\, u_1'(t)dt \right) = -\frac{\pi^5}{(5!)(3!)(32)}x^4 + \text{other terms}, \\
&\vdots
\end{aligned}
$$

$$\tag{76}$$

Considering the first two components $u_0(x)$ and $u_1(x)$, we observe that the two identical terms $\frac{1}{4!}x^4$ appear in these components with opposite signs. Cancelling these terms and justifying that the remaining non-cancelled term of $u_0(x)$ satisfies the given equation yield the exact solution

$$u(x) = \cos x. \tag{77}$$

Notice that we can solve this example by applying the modified decomposition method. We leave it to the reader to use this technique.

Exercises 4.4

Solve the following Fredholm integro-differential equations by using the *Adomian decomposition method, modified decomposition method, or noise terms*

1. $u'(x) = \sinh x + \frac{1}{8}(1 - e^{-1})x - \frac{1}{8}\int_0^1 xtu(t)\,dt, \quad u(0) = 1.$

2. $u'(x) = 1 - \frac{1}{3}x + \int_0^1 xtu(t)\,dt, \quad u(0) = 0.$

3. $u'(x) = xe^x + e^x - x + \int_0^1 xu(t)\,dt, \quad u(0) = 0.$

4. $u'(x) = x \cos x + \sin x - x + \int_0^{\pi/2} xu(t)\, dt, \quad u(0) = 0.$

5. $u''(x) = -\sin x + x - \int_0^{\pi/2} xtu(t)\, dt, \quad u(0) = 0, u'(0) = 1.$

6. $u'''(x) = 6 + x - \int_0^1 xu''(t)\, dt, \quad u(0) = -1, u'(0) = 1, u''(0) = -2.$

7. $u'''(x) = -\cos x + x + \int_0^{\pi/2} xu''(t)\, dt, u(0) = u''(0) = 0, u'(0) = 1.$

8. $u'(x) = \cos x - \sin x - 2x + \dfrac{\pi}{2} + \int_0^{\pi/2} (x - t)u(t)\, dt, u(0) = 1.$

9. $u'(x) = -\sin x - x + \dfrac{\pi}{2} - 1 + \int_0^{\pi/2} (x - t)u(t)\, dt, u(0) = 1.$

10. $u'(x) = \sin x + \cos x + 2 - \dfrac{\pi}{2} + \int_0^{\pi/2} (x - t)u(t)\, dt, u(0) = -1.$

4.5　The Variational Iteration Method

The variational iteration method [14–18] was presented in Chapters 2 and 3 in detail, therefore we introduce a summary of the main steps. The method usually computes the successive approximations that will converge to the exact solution if such a solution exists. Otherwise, the obtained approximations can be sued for numerical computations especially if the problem is concrete and an exact solution cannot be obtained.

　　The standard jth order Fredholm integro-differential equation reads

$$u^{(j)}(x) = f(x) + \int_a^b K(x,t)u(t)dt, \tag{78}$$

where $u^{(j)}(x) = \frac{d^j u}{dx^j}$, and $u(0), u'(0), \cdots, u^{(j-1)}(0)$ are the initial conditions that should be given.

　　The correction functional for the Fredholm integro-differential equation (78) is

$$u_{n+1}(x) = u_n(x) + \int_0^x \lambda(t)\left(u_n^{(i)}(t) - f(t) - \int_a^b K(t,r)u_n(r)\, dr\right) dt. \tag{79}$$

To apply the variational iteration method, we should first determine the Lagrange multiplier $\lambda(t)$ that can be identified optimally as presented in Chapters 2 and 3. Having $\lambda(t)$ determined, an iteration formula should be used for the determination of the determination of the successive approximations $u_{n+1}(x), n \geq 0$ of the solution $u(x)$. The zeroth approximation u_0 can be any selective function, but using the given initial values $u(0), u'(0), \cdots$ are usually used as examined earlier. Consequently, the solution is given by

$$u(x) = \lim_{n \to \infty} u_n(x). \tag{80}$$

It is interesting to recall that the variational iteration method evaluates successive approximations of the solution and not the components of the solution as in ADM. The last obtained approximation can be considered as $u_n(x)$ where its limit should be evaluated as given earlier.

In what follows, the VIM will be examined by investigating the following examples.

Example 1. Use the variational iteration method to solve the Fredholm integro-differential equation

$$u'(x) = 2 - \sin x + \int_0^\pi tu(t)dt, u(0) = 1. \tag{81}$$

The correction functional for this equation is given by

$$u_{n+1}(x) = u_n(x) - \int_0^x \left(u'_n(t) - 2 + \sin t - \int_0^\pi ru_n(r)\,dr \right)\,dt, \tag{82}$$

where we used $\lambda = -1$ for first-order integro-differential equation. Notice that the correction functional involves Volterra and Fredholm integral equations.

We can use the initial condition to select $u_0(x) = u(0) = 1$. Using this $u_0(x) = 1$ into the correction functional gives the following successive approximations

$$
\begin{aligned}
u_0(x) &= 1, \\
u_1(x) &= u_0(x) - \int_0^x \left(u'_0(t) - 2 + \sin t - \int_0^\pi ru_0(r)\,dr \right)\,dt \\
&= \cos x + 2x + \frac{\pi^2}{2}x, \\
u_2(x) &= u_1(x) - \int_0^x \left(u'_0(t) - 2 + \sin t - \int_0^\pi ru_1(r)\,dr \right)\,dt \\
&= (\cos x + 2x) + (-2x + \frac{\pi^2}{2}x) + (-\frac{\pi^2}{2}x + \frac{2\pi}{3}x) \\
&\quad + \quad \text{other terms},
\end{aligned}
\tag{83}
$$

$$u_3(x) \;=\; u_2(x) - \int_0^x \left(u_2'(t) - 2 + \sin t - \int_0^\pi r u_2(r)\, dr \right) dt$$

$$=\; (\cos x + 2x) + \left(-2x + \frac{\pi^2}{2}x\right) + \left(-\frac{\pi^2}{2}x + \frac{2\pi}{3}x\right)$$

$$+\;\; \left(-\frac{2\pi}{3}x + \text{other terms}\right),$$

and so on. The VIM admits the use of

$$u(x) = \lim_{n \to \infty} u_n(x). \tag{84}$$

It is obvious that noise terms appear in the successive approximations, that will be cancelled in the limit. Hence, we obtain the exact solution

$$u(x) = \cos x. \tag{85}$$

Example 2. Use the variational iteration method to solve the Fredholm integro-differential equation

$$u'(x) = \sec^2 x - \ln 2 + \int_0^{\frac{\pi}{3}} u(t)\, dt, \quad u(0) = 0. \tag{86}$$

The correction functional for this equation is given by

$$u_{n+1}(x) = u_n(x) - \int_0^x \left(u_n'(t) - \sec^2 t + \ln 2 - \int_0^{\frac{\pi}{3}} u_n(r)\, dr \right) dt, \tag{87}$$

where we used $\lambda = -1$.

We can use the initial condition to select $u_0(x) = u(0) = 0$. Using this $u_0(x) = 1$ into the correction functional gives the following successive approximations

$$u_0(x) \;=\; 0,$$

$$u_1(x) \;=\; u_0(x) - \int_0^x \left(u_0'(t) - \sec^2 t + \ln 2 - \int_0^{\frac{\pi}{3}} u_0(r)\, dr \right) dt$$

$$=\; \tan x - x \ln 2,$$

$$u_2(x) \;=\; u_1(x) - \int_0^x \left(u_1'(t) - \sec^2 t + \ln 2 - \int_0^{\frac{\pi}{3}} u_1(r)\, dr \right) dt$$

$$= (\tan x - x \ln 2) + (x \ln 2 - \frac{\pi^2 x}{18} \ln 2),$$

$$u_3(x) = u_2(x) - \int_0^x \left(u_2'(t) - \sec^2 t + \ln 2 - \int_0^{\frac{\pi}{3}} u_2(r)\, dr \right) dt$$

$$= (\tan x - x \ln 2) + (x \ln 2 - \frac{\pi^2 x}{18} \ln 2) + (\frac{\pi^2 x}{18} \ln 2 + \cdots),$$

$$(88)$$

and so on. The VIM admits the use of

$$u(x) = \lim_{n \to \infty} u_n(x). \tag{89}$$

It is obvious that noise terms appear in the successive approximations, that will be cancelled in the limit. Hence, we obtain the exact solution

$$u(x) = \tan x. \tag{90}$$

Exercises 4.5

Solve the following Fredholm integro-differential equations by using the *variational iteration method*

1. $u'(x) = \sin x + x \cos x - 1 + \int_0^{\frac{\pi}{2}} u(t)\, dt, \quad u(0) = 0.$

2. $u'(x) = \cos x - x \sin x + 2 + \int_0^{\pi} u(t)\, dt, \quad u(0) = 0.$

3. $u'(x) = 2\sec^2 x \tan x - \frac{\pi^2}{32} + \int_0^{\frac{\pi}{4}} u(t)\, dt, \quad u(0) = 1.$

4. $u'(x) = 3 - 12x + \int_0^1 tu(t)\, dt, \quad u(0) = 1.$

5. $u'(x) = e^x - 1 + \int_0^1 tu(t)\, dt, \quad u(0) = 1.$

6. $u'(x) = -\frac{19}{12} + 2x - 3x^2 + \int_0^1 u(t)\, dt, \quad u(0) = 1.$

4.6   Converting to Fredholm Integral Equations

In this section we will discuss a technique that will convert Fredholm integro-differential equation to an equivalent Fredholm integral equation. This can be easily done by integrating both sides of the integro-differential equation as many times as the order of the derivative involved in the equation from 0 to x for every time we integrate, and by using the given initial conditions.

It is important to note that this technique is applicable only if the Fredholm integro-differential equation involves the unknown function $u(x)$ only, and not any of its derivatives, under the integral sign.

Having established the transformation to a standard Fredholm integral equation, we can use any of the methods that were discussed before in Chapter 2, namely the decomposition method, the variational iteration method, the direct computation method, the successive approximations method or the successive substitutions method.

To give a clear overview of this method we discuss the following illustrative examples.

Example 1.   Solve the following Fredholm integro-differential equation by converting it to a standard Fredholm integral equation

$$u^{'}(x) = 1 - \frac{1}{3}x + x \int_0^1 t\, u(t)dt, \quad u(0) = 0. \tag{91}$$

Integrating both sides from 0 to x and using the initial condition we obtain

$$u(x) = x - \frac{1}{3!}x^2 + \frac{1}{2!}x^2 \left(\int_0^1 t\, u(t)dt \right). \tag{92}$$

It can be easily seen that (92) is a Fredholm integral equation; therefore we can select any method that was introduced earlier. To achieve our goal, we select the successive approximations method to solve this equation. Hence we set a zeroth approximation by

$$u_0(x) = x, \tag{93}$$

and using this choice in (92) yields the first approximation

$$u_1(x) = x - \frac{1}{3!}x^2 + \frac{1}{2!}x^2 \left(\int_0^1 t^2 dt \right), \tag{94}$$

which gives

$$u_1(x) = x. \tag{95}$$

It is obvious that if we continue in the same manner, we then obtain

$$u_n(x) = x. \tag{96}$$

Accordingly,

$$
\begin{aligned}
u(x) &= \lim_{n\to\infty} u_n(x) \\
&= \lim_{n\to\infty} x \tag{97} \\
&= x.
\end{aligned}
$$

Example 2.   Solve the following Fredholm integro-differential equation by converting it to a standard Fredholm integral equation.

$$u''(x) = e^x - x + x \int_0^1 t\,u(t)dt, \quad u(0) = 1, \quad u'(0) = 1. \tag{98}$$

Integrating both sides of (98) twice from 0 to x and using the initial conditions we obtain

$$u(x) = e^x - \frac{1}{3!}x^3 + \frac{1}{3!}x^3 \int_0^1 t u(t)\,dt, \tag{99}$$

a typical Fredholm integral equation. As indicated earlier we can select any method that will determine the solution; therefore we will use the direct computation method for this example. Therefore we can express (99) in the form

$$u(x) = e^x - \frac{1}{3!}x^3 + \alpha\frac{1}{3!}x^3, \tag{100}$$

where the constant α is defined by the definite integral

$$\alpha = \int_0^1 t u(t)\,dt. \tag{101}$$

Substituting (100) into (101) we obtain

$$\alpha = \int_0^1 t \left(e^t - \frac{1}{3!}t^3 + \alpha\frac{1}{3!}t^3 \right) dt, \tag{102}$$

an easy integral to evaluate, from which we obtain

$$\alpha = 1. \tag{103}$$

Inserting the value of α obtained in (103) into (100) yields the exact solution given by

$$u(x) = e^x. \tag{104}$$

Exercises 4.6

Solve the following Fredholm integro-differential equations by *converting it to Fredholm integral equations*

1. $u'(x) = -x \sin x + \cos x + (1 - \pi/2)x + \int_0^{\pi/2} xu(t)\, dt, \quad u(0) = 0.$

2. $u''(x) = -e^x + \dfrac{1}{2}x + \int_0^1 xtu(t)\, dt, \quad u(0) = 0, u'(0) = -1.$

3. $u''(x) = -\sin x + \cos x + (2 - \pi/2)x - \int_0^{\pi/2} xtu(t)\, dt,$

$u(0) = -1, u'(0) = 1.$

4. $u'(x) = \dfrac{7}{6} - 11x - \int_0^1 (x - t)u(t)\, dt, \quad u(0) = 0.$

5. $u'(x) = \dfrac{1}{4} + \cos(2x) - \int_0^{\pi/4} xu(t)\, dt, \quad u(0) = 0.$

Chapter 5

Volterra Integro-Differential Equations

5.1 Introduction

In this chapter we will be concerned with the Volterra integro-differential equations where both differential and integral operators will appear in the same equation. This style of equations was introduced by Volterra for the first time in the early 1900. Volterra was investigating a population growth model, focusing his study on the hereditary influences, where through his research work the topic of integro-differential equations was established. Scientists and researchers investigated the topic of integro-differential equations through their work in science applications such as heat transfer, diffusion process in general, neutron diffusion and biological species coexisting together with increasing and decreasing rates of generating. More details about the sources where these equations arise can be found in physics, biology and engineering applications as well as in advanced integral equations books.

In the integro-differential equations, it is important to note that the unknown function $u(x)$ and one or more of its derivatives such as $u'(x)$, $u''(x)$, ..., appear outside and inside the integral sign as well. One quick source of integro-differential equations can be clearly seen when we convert a differential equation to an integral equation using Leibniz rule. The Volterra integro-differential equation can be viewed in this case as an intermediate stage when finding an equivalent Volterra integral equation to

145

the given differential equation as discussed in Section 1.5.

The following are examples of linear Volterra integro-differential equations:

$$u^{'}(x) \;=\; x - \int_0^x (x-t)u(t)dt, \quad u(0) = 0, \tag{1}$$

$$u^{''}(x) \;=\; -x + \int_0^x (x-t)u(t)dt, \quad u(0) = 0, u^{'}(0) = -1. \tag{2}$$

It is clear from the examples given above that the unknown function $u(x)$ or one of its derivatives appear under the integral sign, and other derivatives of $u(x)$ appear outside the integral sign as well. Therefore, the above given equations involve the derivatives and the integral operators in the same equation, and consequently the term integro-differential equations has been used for problems involving this combination of operators.

Examining the limits of integrals in equations (1)-(2) and following the classification concept used in Chapter 1 allow us to use the classification *Volterra integro-differential equations* to equations (1) and (2). In addition, it is also interesting to know that equations (1)-(2) are *linear* Volterra integro-differential equations, and this is related to the linearity occurrence of the unknown function $u(x)$ and its derivatives in the equations above. However, *nonlinear* integro-differential equations also arise in many scientific and engineering problems. Our concern in this chapter will be focused on the *linear* Volterra integro-differential equations.

To determine a solution for the Volterra integro-differential equation, the initial conditions should be given, and this may be clearly seen as a result of involving $u(x)$ and its derivatives. The initial conditions are needed to determine the constants of integration.

5.2   Volterra Integro-Differential Equations

In this section we will present the reliable methods that will be used to handle Volterra integro-differential equations. This new type of equations was termed as Volterra integro-differential equations, given in the form

$$u^{(n)}(x) = f(x) + \lambda \int_0^x K(x,t)u(t)dt. \tag{3}$$

We will focus our study on equations that involve separable kernels of the form

$$K(x,t) = \sum_{k=1}^n g_k(x)\, h_k(t). \tag{4}$$

Without loss of generality, we will consider the cases where the kernel $K(x,t)$ consists of one product of the functions $g(x)$ and $h(t)$ given by

$$K(x,t) = g(x)\,h(t), \tag{5}$$

where other cases can be generalized in the same manner. The nonseparable kernel can be reduced to separable kernel by using the Taylor expansion for the kernel involved. The methods to be introduced are identical, with some exceptions, to the methods discussed in Chapter 3. Our approach will be mainly based on how we can extend the methods used in Chapter 3 to handle this type of equations. For this reason we first start with the most practical method.

5.3 The Series Solution Method

This method has been extensively introduced in Chapter 3. Without loss of generality, we may consider a standard form to the Volterra integro-differential equation given by

$$u^{(n)}(x) = f(x) + \int_0^x K(x,t)\,u(t)dt, \quad u^{(k)}(0) = b_k, \; 0 \le k \le (n-1), \tag{6}$$

where $u^{(n)}(x)$ indicates the nth derivative of $u(x)$ with respect to x, and b_k are constants that define the initial conditions. Substituting (5) into (6) yields

$$u^{(n)}(x) = f(x) + g(x) \int_0^x h(t)\,u(t)dt, \quad u^{(k)}(0) = b_k, \; 0 \le k \le (n-1). \tag{7}$$

We will follow a manner parallel to the approach of the *series solution method* that usually used in solving ordinary differential equations around an ordinary point. To achieve this goal, we first assume that the solution $u(x)$ of (7) is an analytic function and hence can be represented by a series expansion given by

$$u(x) = \sum_{k=0}^{\infty} a_k x^k, \tag{8}$$

where the coefficients a_k are constants that will be determined. It is to be noted that the first few coefficients a_k can be determined by using the initial conditions so that

$$
\begin{aligned}
a_0 &= u(0), \\
a_1 &= u'(0), \\
a_2 &= \tfrac{1}{2!}\,u''(0),
\end{aligned}
\tag{9}
$$

and so on depending on the number of the initial conditions given, whereas the remaining coefficients a_k will be determined from applying the technique as will be discussed later. Substituting (8) into both sides of (7) yields

$$\left(\sum_{k=0}^{\infty} a_k x^k\right)^{(n)} = f(x) + g(x) \int_0^x h(t) \left(\sum_{k=0}^{\infty} a_k t^k\right) dt. \tag{10}$$

In view of (10), the integral equation (7) will be reduced to several calculable integrals in the right hand side of (10) that can be easily evaluated where we have to integrate terms of the form t^n, $n \geq 0$ only.

The next step is to write the Taylor expansion for $f(x)$, evaluate the resulting traditional integrals in (10), and then equating the coefficients of like powers of x in both sides of the equation. Accordingly, this leads to a complete determination of the coefficients $a_i, i \geq 0$.

Consequently, substituting the obtained coefficients a_k, $k \geq 0$ into (8) produces the solution in a series form. This may converge to a solution in a closed form, if an exact solution exists, or we may use the obtained series for numerical purposes.

To give a clear overview of the technique and how it should be implemented for Volterra integro-differential equations, the series solution method will be illustrated by discussing the following examples.

Example 1. Solve the following Volterra integro-differential equation by using the series solution method.

$$u''(x) = x \cosh x - \int_0^x t\, u(t)\, dt, \quad u(0) = 0, u'(0) = 1. \tag{11}$$

Substituting $u(x)$ by the series

$$u(x) = \sum_{n=0}^{\infty} a_n x^n, \tag{12}$$

into both sides of the equation (11) and using the Taylor expansion of $\cosh x$ we obtain

$$\sum_{n=2}^{\infty} n(n-1)a_n x^{n-2} = x \left(\sum_{k=0}^{\infty} \frac{x^{2k}}{(2k)!}\right) - \int_0^x t \left(\sum_{n=0}^{\infty} a_n t^n\right) dt. \tag{13}$$

Using the initial conditions yields

$$a_0 = 0, \tag{14}$$

$$a_1 = 1. \tag{15}$$

Evaluating the traditional integrals that involve terms of the form t^n, $n \geq 0$, and using few terms from both sides yield

$$2a_2 + 6a_3 x + 12a_4 x^2 + 20a_5 x^3 + \cdots = x\left(1 + \frac{1}{2!}x^2 + \frac{1}{4!}x^4 + \cdots\right)$$

$$-\left(\frac{1}{3}x^3 + \frac{1}{4}a_2 x^4 + \cdots\right). \tag{16}$$

Equating the coefficients of like powers of x in both sides we find

$$a_2 = 0,$$

$$a_3 = \frac{1}{3!} \tag{17}$$

$$a_4 = 0,$$

and generally

$$a_{2n} = 0, \quad \text{for} \quad n \geq 0, \tag{18}$$

and

$$a_{2n+1} = \frac{1}{(2n+1)!}, \quad \text{for} \quad n \geq 0. \tag{19}$$

Using (12) we find the solution $u(x)$ in a series form

$$u(x) = x + \frac{1}{3!}x^3 + \frac{1}{5!}x^5 + \frac{1}{7!}x^7 + \cdots, \tag{20}$$

and in a closed form, the exact solution is given by

$$u(x) = \sinh x, \tag{21}$$

obtained upon using the Taylor series of $\sinh x$.

Example 2. As a second example we use the series solution method to solve

$$u''(x) = \cosh x + \frac{1}{4} - \frac{1}{4}\cosh 2x + \int_0^x \sinh tu(t)dt, \quad u(0) = 1,\, u'(0) = 0. \tag{22}$$

Using (8) and considering the first few terms of the expansion of $u(x)$, we obtain

$$u(x) = 1 + a_2 x^2 + a_3 x^3 + a_4 x^4 + a_5 x^5 + \cdots. \tag{23}$$

Substituting (23) into both sides of (22) yields

$$2a_2 + 6a_3x + 12a_4x^2 + 20a_5x^3 + \cdots = \left(1 + \frac{x^2}{2!} + \frac{x^4}{4!} + \cdots\right) + \frac{1}{4}$$

$$-\frac{1}{4}\left(1 + \frac{(2x)^2}{2!} + \frac{(2x)^4}{4!} + \cdots\right)$$

$$+\int_0^x \left(t + \frac{t^3}{3!} + \frac{t^5}{5!} + \cdots\right)\left(1 + a_2t^2 + a_3t^3 + \cdots\right) dt. \quad (24)$$

Integrating the right hand side and equating the coefficients of like powers of x we find

$$\begin{aligned}
a_0 &= 1, \\
a_1 &= 0, \\
a_2 &= \tfrac{1}{2!}, \\
a_3 &= 0, \\
a_4 &= \tfrac{1}{4!}, \\
a_5 &= 0,
\end{aligned} \qquad (25)$$

and so on, where the constants a_0 and a_1 are defined by using the initial conditions. Consequently the solution in a series form is given by

$$u(x) = 1 + \frac{x^2}{2!} + \frac{x^4}{4!} + \frac{x^6}{6!} + \cdots, \quad (26)$$

which gives the exact solution in a closed form

$$u(x) = \cosh x. \quad (27)$$

Exercises 5.3

Solve the following Volterra integro-differential equations by using *the series solution method*

1. $u^{'}(x) = 1 - 2x\sin x + \displaystyle\int_0^x u(t)\, dt, \quad u(0) = 0.$

2. $u^{'}(x) = -1 + \dfrac{1}{2}x^2 - xe^x - \displaystyle\int_0^x tu(t)\, dt, \quad u(0) = 0.$

3. $u^{''}(x) = 1 - x(\cos x + \sin x) - \displaystyle\int_0^x tu(t)\, dt, \quad u(0) = -1, u^{'}(0) = 1.$

4. $u^{''}(x) = -8 - \dfrac{1}{3}(x^3 - x^4) + \displaystyle\int_0^x (x - t)u(t)\, dt, \quad u(0) = 0, u^{'}(0) = 2.$

5. $u^{''}(x) = \dfrac{1}{2}x^2 - x\cosh x - \displaystyle\int_0^x tu(t)\, dt, \quad u(0) = 1, u^{'}(0) = -1.$

5.4   The Adomian Decomposition Method

The Adomian decomposition method (ADM) and the modified decomposition method were discussed in detail in Chapters 2, 3 and 4. In this section we will introduce how this successful method can be implemented to determine a series solution to the Volterra integro-differential equations. Without loss of generality, we may assume a standard form to the Volterra integro-differential equation defined by the standard form

$$u^{(n)}(x) = f(x) + \int_0^x K(x,t)\, u(t)dt, \quad u^{(k)}(0) = b_k, \, 0 \le k \le (n-1) \quad (28)$$

where $u^{(n)}(x)$ indicates the nth derivative of $u(x)$ with respect to x and b_k are constants that define the initial conditions. It is natural to seek an expression for $u(x)$ that will be derived from (28). This can be done by integrating both sides of (28) from 0 to x as many times as the order of the derivative involved. Consequently, we obtain

$$u(x) = \sum_{k=0}^{n-1} \frac{1}{k!} b_k x^k + L^{-1}\left(f(x)\right) + L^{-1}\left(\int_0^x K(x,t)\, u(t)dt\right), \quad (29)$$

where $\sum_{k=0}^{n-1} \frac{1}{k!} b_k x^k$ is obtained by using the initial conditions, and L^{-1} is an n-fold integration operator. Now we apply the decomposition method by defining the solution $u(x)$ of (29) in a decomposition series given by

$$u(x) = \sum_{n=0}^{\infty} u_n(x). \quad (30)$$

Substituting (30) into both sides of (29) we get

$$\sum_{n=0}^{\infty} u_n(x) = \sum_{k=0}^{n-1} \frac{1}{k!} b_k x^k + L^{-1}\left(f(x)\right) + L^{-1}\left(\int_0^x K(x,t)\left(\sum_{n=0}^{\infty} u_n(t)\right) dt\right),$$
$$(31)$$

or equivalently

$$u_0(x) + u_1(x) + u_2(x) + \cdots = \sum_{k=0}^{n-1} \frac{1}{k!} b_k x^k + L^{-1}\left(f(x)\right)$$

$$+ L^{-1} \left(\int_0^x K(x,t)\, u_0(t) dt \right)$$

$$+ L^{-1} \left(\int_0^x K(x,t)\, u_1(t) dt \right)$$

$$+ L^{-1} \left(\int_0^x K(x,t)\, u_2(t) dt \right) \tag{32}$$

$$+ \cdots .$$

The components $u_i(x), i \geq 0$ of the unknown function $u(x)$ are determined in a recursive manner, in a similar way as discussed before, if we set

$$u_0(x) = \sum_{k=0}^{n-1} \frac{1}{k!} a_k x^k + L^{-1}\left(f(x) \right), \tag{33}$$

$$u_1(x) = L^{-1} \left(\int_0^x K(x,t)\, u_0(t) dt \right), \tag{34}$$

$$u_2(x) = L^{-1} \left(\int_0^x K(x,t)\, u_1(t) dt \right), \tag{35}$$

$$u_3(x) = L^{-1} \left(\int_0^x K(x,t)\, u_2(t) dt \right), \tag{36}$$

and so on. The decomposition method discussed above for the determination of the components $u_i(x), i \geq 0$ of the solution $u(x)$ of the equation (28) can be written in a recursive manner by

$$u_0(x) = \sum_{k=0}^{n-1} \frac{1}{k!} a_k x^k + L^{-1}\left(f(x) \right) \tag{37}$$

$$u_{n+1}(x) = L^{-1} \left(\int_0^x K(x,t)\, u_n(t) dt \right), \quad n \geq 0. \tag{38}$$

In view of (37) and (38), the components $u_i(x), i \geq 0$ are immediately evaluated. With the components determined, the solution $u(x)$ of (28) is then obtained in a series form using (30). Consequently, the series obtained for $u(x)$ mostly provides the exact solution in a closed form as will be illustrated later. However, for concrete problems, where (30) cannot be evaluated, a truncated series $\sum_{n=0}^{k} u_n(x)$ is usually used to approximate the solution $u(x)$.

It is convenient to point out that the phenomenon of the self-cancelling noise terms that was introduced before may be applied here if the noise terms appear between $u_0(x)$ and $u_1(x)$. Moreover, the modified decomposition method can be used as well. The following examples will explain how we can use the Adomian decomposition technique.

Example 1. Solve the following Volterra integro-differential equation by using the decomposition method.

$$u''(x) = x + \int_0^x (x - t)\, u(t)dt, \quad u(0) = 0, u'(0) = 1. \tag{39}$$

Applying the *two-fold* integral operator L^{-1}

$$L^{-1}(.) = \int_0^x \int_0^x (.)dx\, dx, \tag{40}$$

to both sides of (39), i.e. integrating both sides of (39) twice from 0 to x, and using the given initial conditions yield

$$u(x) = x + \frac{1}{3!}x^3 + L^{-1}\left(\int_0^x (x - t)\, u(t)dt\right). \tag{41}$$

Following the decomposition scheme (37) and (38) we find

$$
\begin{aligned}
u_0(x) &= x + \tfrac{1}{3!}x^3, \\
u_1(x) &= L^{-1}\left(\int_0^x (x - t)\, u_0(t)dt\right), \\
&= \tfrac{1}{5!}x^5 + \tfrac{1}{7!}x^7, \\
u_2(x) &= L^{-1}\left(\int_0^x (x - t)\, u_1(t)dt\right), \\
&= \frac{1}{9!}x^9 + \frac{1}{11!}x^{11}, \\
u_3(x) &= L^{-1}\left(\int_0^x (x - t)\, u_2(t)dt\right), \\
&= \frac{1}{13!}x^{13} + \frac{1}{15!}x^{15}, \\
&\vdots
\end{aligned}
\tag{42}
$$

Combining the last results yields the solution $u(x)$ in a series form given by

$$u(x) = x + \frac{1}{3!}x^3 + \frac{1}{5!}x^5 + \frac{1}{7!}x^7 + \frac{1}{9!}x^9 + \frac{1}{11!}x^{11} + \frac{1}{13!}x^{13} + \frac{1}{15!}x^{15} + \cdots, \tag{43}$$

and this leads to the exact solution in a closed form

$$u(x) = \sinh x. \tag{44}$$

Example 2.   Solve the following Volterra integro-differential equation

$$u''(x) = 1 + \int_0^x (x - t)\, u(t)dt, \quad u(0) = 1, u'(0) = 0, \tag{45}$$

by using the decomposition method. Integrating both sides of (45) twice from 0 to x and using the given initial conditions yield

$$u(x) = 1 + \frac{1}{2!}x^2 + L^{-1}\left(\int_0^x (x - t)\, u(t)dt\right), \tag{46}$$

where L^{-1} is a two-fold integration operator given above by (40).
   Following the decomposition method we obtain

$$
\begin{aligned}
u_0(x) &= 1 + \tfrac{1}{2!}x^2, \\[2mm]
u_1(x) &= L^{-1}\left(\int_0^x (x - t)\, u_0(t)dt\right), \\[2mm]
&= \frac{1}{4!}x^4 + \frac{1}{6!}x^6, \\[2mm]
u_2(x) &= L^{-1}\left(\int_0^x (x - t)\, u_1(t)dt\right), \\[2mm]
&= \frac{1}{8!}x^8 + \frac{1}{10!}x^{10},
\end{aligned}
\tag{47}
$$

$$\vdots$$

Combining the obtained results yields the solution $u(x)$ in a series form given by

$$u(x) = 1 + \frac{1}{2!}x^2 + \frac{1}{4!}x^4 + \frac{1}{6!}x^6 + \frac{1}{8!}x^8 + \frac{1}{10!}x^{10} + \cdots, \tag{48}$$

and this gives the exact solution in a closed form

$$u(x) = \cosh x. \tag{49}$$

Example 3.   Solve the following Volterra integro-differential equation

$$u'''(x) = -1 + \int_0^x u(t)dt, \quad u(0) = u'(0) = 1, u''(0) = -1 \tag{50}$$

by using the decomposition method.

We note here that the Volterra integro-differential equation involves the third order differential operator $u'''(x)$, therefore integrating both sides of (50) three times from 0 to x and using the initial conditions we obtain

$$u(x) = 1 + x - \frac{1}{2!}x^2 - \frac{1}{3!}x^3 + L^{-1}\left(\int_0^x u(t)dt\right). \qquad (51)$$

Following the decomposition scheme we find

$$u_0(x) = 1 + x - \frac{1}{2!}x^2 - \frac{1}{3!}x^3, \qquad (52)$$

which gives

$$u_1(x) = L^{-1}\left(\int_0^x u_0(t)dt\right)$$
$$= \frac{1}{4!}x^4 + \frac{1}{5!}x^5 - \frac{1}{6!}x^6 - \frac{1}{7!}x^7. \qquad (53)$$

Consequently, the solution $u(x)$ given in a series form

$$u(x) = 1 + x - \frac{1}{2!}x^2 - \frac{1}{3!}x^3 + \frac{1}{4!}x^4 + \frac{1}{5!}x^5 - \frac{1}{6!}x^6 - \frac{1}{7!}x^7 + \cdots \qquad (54)$$

We can easily observe that the series solution obtained in (54) will not easily give the closed form solution; however, rewriting (54) by

$$u(x) = \left(1 - \frac{1}{2!}x^2 + \frac{1}{4!}x^4 + \cdots\right) + \left(x - \frac{1}{3!}x^3 + \frac{1}{5!}x^5 + \cdots\right), \qquad (55)$$

provides the closed form solution given by

$$u(x) = \cos x + \sin x. \qquad (56)$$

Exercises 5.4

Solve the following Volterra integro-differential equations by using the *Adomian decomposition method*

1. $u''(x) = 1 + x - \frac{1}{3!}x^3 + \int_0^x (x-t)u(t)\, dt, \quad u(0) = 1, u'(0) = 2.$

2. $u''(x) = -1 - \frac{1}{2!}x^2 + \int_0^x (x-t)u(t)\, dt, \quad u(0) = 2, u'(0) = 0.$

3. $u'(x) = 2 + \int_0^x u(t)\, dt, \quad u(0) = 2.$

4. $u'(x) = 1 - \int_0^x u(t)\, dt, \quad u(0) = 1.$

5. $u^{(iv)}(x) = -x + \frac{1}{2!}x^2 - \int_0^x (x-t)u(t)\, dt,$
$u(0) = 1, u'(0) = -1, u''(0) = 0, u'''(0) = 1.$

5.5   The Variational Iteration Method

The variational iteration method was used effectively in the preceding chapters. This method was used to handle both Fredholm and Volterra integral equations. The method was presented before where we studied the necessary issues that should be addressed. We should first give the correction functional that works for differential equation provided that the Lagrange multiplier is derived. The method gives successive approximations of the exact solution and does not give distinct components as in the case of Adomian decomposition method. Because we will study the Volterra integro-differential equations in this section, then we can apply this method directly, and we do not need any specific condition. In what follows we give a brief summary to the essential steps of the variational method.

The standard jth order integro-differential equation is of the form

$$u^{(j)}(x) = f(x) + \int_0^x K(x,t)u(t)dt, \tag{57}$$

where $u^{(j)}(x) = \frac{d^j u}{dx^j}$, and $u(0), u'(0), \cdots, u^{(j-1)}(0)$ are the initial conditions.

The correction functional for the integro-differential equation (57) is

$$u_{n+1}(x) = u_n(x) + \int_0^x \lambda(x,t) \left(u_n^{(i)}(t) - f(t) - \int_0^t K(t,r)u_n(r)\,dr \right) dt. \tag{58}$$

Recall that the Lagrange multipliers were identified optimally, as shown before, and summarized by

$$\begin{aligned}
&u' + f(u(\xi), u'(\xi)) = 0, \lambda = -1, \\
&u'' + f(u(\xi), u'(\xi), u''(\xi)) = 0, \lambda = \xi - x, \\
&u''' + f(u(\xi), u'(\xi), u''(\xi), u'''(\xi)) = 0, \lambda = -\frac{1}{2!}(\xi - x)^2,
\end{aligned} \tag{59}$$

and so on. Moreover, the zeroth approximation $u_0(x)$ can be preferably selected by using the initial values $u(0), u'(0), \dots$ and Taylor series. Having determined the first few successive approximations, the solution is given by

$$u(x) = \lim_{n \to \infty} u_n(x). \tag{60}$$

The VIM will be illustrated by studying the following examples.

Example 1. Use the variational iteration method to solve the Volterra integro-differential equation

$$u'(x) = -x + \frac{1}{2}x^2 + \int_0^x u(t)dt, u(0) = 2. \tag{61}$$

The correction functional for this equation is given by

$$u_{n+1}(x) = u_n(x) - \int_0^x \left(u_n'(t) + t - \frac{1}{2}t^2 - \int_0^t u_n(r)\, dr \right) dt, \qquad (62)$$

where we used $\lambda = -1$ for first-order integro-differential equations.

We can use the initial condition to select $u_0(x) = u(0) = 2$. Using this selection into the correction functional gives the following successive approximations

$$
\begin{aligned}
u_0(x) &= 2, \\
u_1(x) &= u_0(x) - \int_0^x \left(u_0'(t) + t - \frac{1}{2}t^2 - \int_0^t u_0(r)\, dr \right) dt, \\
&= 2 + \frac{1}{2!}x^2 + \frac{1}{3!}x^3, \\
u_2(x) &= u_1(x) - \int_0^x \left(u_1'(t) + t - \frac{1}{2}t^2 - \int_0^t u_1(r)\, dr \right) dt, \qquad (63) \\
&= 2 + \frac{1}{2!}x^2 + \frac{1}{3!}x^3 + \frac{1}{4!}x^4 + \frac{1}{5!}x^5, \\
u_3(x) &= u_2(x) - \int_0^x \left(u_2'(t) + t - \frac{1}{2}t^2 - \int_0^t u_2(r)\, dr \right) dt, \\
&= 2 + \frac{1}{2!}x^2 + \frac{1}{3!}x^3 + \frac{1}{4!}x^4 + \frac{1}{5!}x^5 + \frac{1}{6!}x^6 + \frac{1}{7!}x^7,
\end{aligned}
$$

and so on. The VIM admits the use of

$$u(x) = \lim_{n \to \infty} u_n(x), \qquad (64)$$

that gives the exact solution

$$u(x) = 1 - x + e^x. \qquad (65)$$

Example 2. Use the variational iteration method to solve the Volterra integro-differential equation

$$u'(x) = 3e^x - x - 2 + \int_0^x (x - t)u(t)dt,\ u(0) = 0. \qquad (66)$$

The correction functional for this equation is given by

$$u_{n+1}(x) = u_n(x) - \int_0^x \left(u_n'(t) - 3e^t + t + 2 - \int_0^t (t - r)u_n(r)\, dr \right) dt, \qquad (67)$$

where we used $\lambda = -1$ for first-order integro-differential equations.

We can use the initial condition to select $u_0(x) = u(0) = 2$. Using this selection into the correction functional gives the following successive approximations

$$u_0(x) = 0,$$

$$u_1(x) = u_0(x) - \int_0^x \left(u_0'(t) - 3e^t + t + 2 - \int_0^t (t-r)u_0(r)\, dr \right)\, dt$$

$$= x + x^2 + \tfrac{1}{2}x^3 + \tfrac{1}{8}x^4 + \tfrac{1}{40}x^5 + \tfrac{1}{240}x^6 + \cdots,$$

$$u_2(x) = u_0(x) - \int_0^x \left(u_1'(t) - 3e^t + t + 2 - \int_0^t (t-r)u_1(r)\, dr \right)\, dt$$

$$= x + x^2 + \tfrac{1}{2!}x^3 + \tfrac{1}{3!}x^4 + \tfrac{1}{4!}x^5 + \tfrac{1}{5!}x^6 + \cdots,$$

$$(68)$$

and so on. The VIM admits the use of

$$u(x) = \lim_{n\to\infty} u_n(x), \tag{69}$$

that gives the exact solution

$$u(x) = xe^x. \tag{70}$$

Example 3. Use the variational iteration method to solve the Volterra integro-differential equation

$$u''(x) = -x + \int_0^x (x-t)u(t)dt, \, u(0) = 0, \, u'(0) = 1. \tag{71}$$

The correction functional for this equation is given by

$$u_{n+1}(x) = u_n(x) + \int_0^x (t-x)\left(u_n''(t) + t - \int_0^t (t-r)u_n(r)\, dr \right)\, dt, \tag{72}$$

where we used $\lambda = (t-x)$ for second-order integro-differential equations. We can use the initial condition to select $u_0(x) = u(0) + xu'(0) = x$. Using this selection into the correction functional gives the following successive approximations

$$u_0(x) = x,$$

$$u_1(x) = u_0(x) + \int_0^x (t-x)\left(u_0''(t) + t - \int_0^t (t-r)u_0(r)\, dr \right)\, dt$$

$$= x - \tfrac{1}{3!}x^3 + \tfrac{1}{5!}x^5,$$

$$(73)$$

$$u_2(x) = u_1(x) + \int_0^x (t-x)\left(u_1''(t) + t - \int_0^t (t-r)u_1(r)\, dr \right)\, dt$$

$$= x - \tfrac{1}{3!}x^3 + \tfrac{1}{5!}x^5 - \tfrac{1}{7!}x^7 + \tfrac{1}{9!}x^9,$$

and so on. The VIM admits the use of

$$u(x) = \lim_{n\to\infty} u_n(x), \tag{74}$$

that gives the exact solution

$$u(x) = \sin x. \tag{75}$$

Exercises 5.5

Solve the following Volterra integro-differential equations by using the *variational iteration method*

1. $u^{'}(x) = 2e^x - 1 + \int_0^x u(t)\,dt, \quad u(0) = 0.$

2. $u^{'}(x) = 2\cos x - \frac{1}{2}x^2 + \int_0^x u(t)\,dt, \quad u(0) = 0.$

3. $u^{'}(x) = 1 + 2\sin x - \frac{1}{2}x^2 + \int_0^x u(t)\,dt, \quad u(0) = -1.$

4. $u^{'}(x) = 1 - \frac{1}{2}x^2 + \int_0^x u(t)\,dt, \quad u(0) = 1.$

5. $u^{''}(x) = 1 + x - \frac{1}{6}x^3 + \int_0^x (x-t)u(t)\,dt, \quad u(0) = 1, u'(0) = 2.$

6. $u^{''}(x) = x - 1 + \int_0^x (x-t)u(t)\,dt, \quad u(0) = 1, u'(0) = -1.$

5.6 Converting to Volterra Integral Equation

We can easily convert the Volterra integro-differential equation to an equivalent Volterra integral equation, provided that the kernel is a *difference kernel* defined by the form $K(x,t) = K(x-t)$. This can be easily done by integrating both sides of the equation and using the initial conditions. To perform the conversion to a standard Volterra integral equation we should use the formula (70) of Chapter 1 that converts multiple integral to a single integral. The reader is advised to review that formula for further reference. The following two specific formulas

$$\int_0^x \int_0^x u(t)\,dt\,dt = \int_0^x (x-t)u(t)\,dt, \tag{76}$$

and

$$\int_0^x \int_0^x \int_0^x u(t)\,dt\,dt\,dt = \frac{1}{2!}\int_0^x (x-t)^2 u(t)\,dt, \tag{77}$$

given by (71) and (72) in Chapter 1 are usually used to transform double integrals and triple integrals respectively to a single integral. Having established the transformation to a standard Volterra integral equation, we may proceed using any of the alternative methods that were discussed before in Chapter 3. To give a clear overview of this method we discuss the following examples.

Example 1. Solve the following Volterra integro-differential equation

$$u'(x) = 2 - \frac{1}{4}x^2 + \frac{1}{4}\int_0^x u(t)dt, \quad u(0) = 0, \tag{78}$$

by converting to a standard Volterra integral equation.

Integrating both sides from 0 to x and using the initial condition we obtain

$$u(x) = 2x - \frac{1}{12}x^3 + \frac{1}{4}\int_0^x \int_0^x u(t)\, dt\, dt, \tag{79}$$

which gives

$$u(x) = 2x - \frac{1}{12}x^3 + \frac{1}{4}\int_0^x (x - t)\, u(t)\, dt, \tag{80}$$

upon using the formula (76). It is clearly seen that (80) is a standard Volterra integral equation that will be solved by using the decomposition method. Following the Adomian decomposition method we set

$$u_0(x) = 2x - \frac{1}{12}x^3, \tag{81}$$

which gives

$$u_1(x) = \frac{1}{4}\int_0^x (x - t)\left(2t - \frac{1}{12}t^3\right)\, dt, \tag{82}$$

so that

$$u_1(x) = \frac{1}{12}x^3 - \frac{1}{240}x^5. \tag{83}$$

We can easily observe that the noise terms $\pm\frac{1}{12}x^3$ appear in the components $u_0(x)$ and $u_1(x)$, and by cancelling this noise term from $u_0(x)$ and justifying that

$$u(x) = 2x, \tag{84}$$

is the exact solution of (80).

Example 2. Solve the following Volterra integro-differential equation

$$u''(x) = -1 + \int_0^x (x - t)u(t)dt, \quad u(0) = 1, u'(0) = 0, \tag{85}$$

by converting to a standard Volterra integral equation.

Integrating both sides twice from 0 to x and using the initial conditions we obtain

$$\begin{aligned}
u'(x) &= -x + \int_0^x \int_0^x (x - t)\, u(t)\, dt\, dt \\
&= -x + \frac{1}{2!}\int_0^x (x - t)^2 u(t)dt,
\end{aligned} \tag{86}$$

by using formula (76), and

$$u(x) = 1 - \frac{1}{2!}x^2 + \int_0^x \int_0^x \int_0^x (x-t)\,u(t)\,dt\,dt\,dt$$
$$= 1 - \frac{1}{2!}x^2 + \frac{1}{3!}\int_0^x (x-t)^3 u(t)dt, \tag{87}$$

by using formula (77). The last equation is a standard Volterra integral equation that will be solved by using the modified decomposition method. To determine $u(x)$ we set

$$u_0(x) = 1, \tag{88}$$

which gives

$$u_1(x) = -\frac{1}{2!}x^2 + \frac{1}{4!}x^4. \tag{89}$$

We can easily observe that the noise terms did not appear between the components $u_0(x)$ and $u_1(x)$, therefore we continue to find more terms to study the solution more closely. Consequently, we find

$$u_2(x) = -\frac{1}{6!}x^6 + \frac{1}{8!}x^8. \tag{90}$$

Combining the results for the components $u_0(x), u_1(x)$ and $u_2(x)$ we obtain the series expression for the solution given by

$$u(x) = 1 - \frac{1}{2!}x^2 + \frac{1}{4!}x^4 - \frac{1}{6!}x^6 + \frac{1}{8!}x^8 + \cdots, \tag{91}$$

which gives the exact solution

$$u(x) = \cos x. \tag{92}$$

Exercises 5.6

Solve the following Volterra integro-differential equations by converting the problem to *Volterra integral equation*

1. $u''(x) = 1 + \int_0^x (x-t)u(t)\,dt, \quad u(0) = 1, u'(0) = 0.$

2. $u'(x) = 1 - \int_0^x u(t)\,dt, \quad u(0) = 0.$

3. $u''(x) = x + \int_0^x (x-t)u(t)\,dt, \quad u(0) = 0, u'(0) = 1.$

4. $u'(x) = 2 - \frac{1}{2!}x^2 + \int_0^x u(t)\,dt, \quad u(0) = 1.$

5. $u'(x) = 1 - \int_0^x u(t)\,dt, \quad u(0) = 1.$

6. $u''(x) = 1 + x + \int_0^x (x-t)u(t)\,dt, \quad u(0) = u'(0) = 1.$

5.7 Converting to Initial Value Problems

In this section we will study how to convert the Volterra integro-differential equation to an equivalent initial value problem, focusing our discussion on the case where the kernel is a difference kernel where $K(x,t) = K(x - t)$. This can be achieved easily by differentiating both sides of the integro-differential equation as many times as needed to remove the integral sign. In differentiating the integral involved we shall use the Leibniz rule to achieve our goal. The Leibniz rule has been extensively introduced in Section 1.4. It is important to note that we should define the initial conditions at every step of differentiation. A similar technique was discussed and examined in Chapters 1 and 2. The reader is advised to review the related material for further use.

Having converted the Volterra integro-differential equation to an initial value problem, the various methods that are used in any ordinary differential equation course can be used to determine the solution. The idea is easy to use but requires more calculations if compared with the integral equations techniques.

To give a clear overview of this method we discuss the following illustrative examples.

Example 1. Solve the following Volterra integro-differential equation by converting it to an initial value problem.

$$u'(x) = 1 + \int_0^x u(t)dt, \quad u(0) = 0. \tag{93}$$

Differentiating both sides of (93) with respect to x and using the Leibniz rule to differentiate the integral at the right hand side we obtain

$$u''(x) = u(x), \tag{94}$$

with initial conditions given by

$$u(0) = 0, \quad u'(0) = 1, \tag{95}$$

where the last initial condition was obtained by substituting $x = 0$ in both sides of (93). The characteristic equation of (93) is

$$r^2 - 1 = 0, \tag{96}$$

which gives the roots

$$r = \pm 1 \tag{97}$$

so that the general solution is given by

$$u(x) = A \cosh x + B \sinh x, \tag{98}$$

where A and B are constants to be determined. Using the initial conditions given by (95) to find the constants A and B, we find that the solution is

$$u(x) = \sinh x. \tag{99}$$

Example 2. Solve the following Volterra integro-differential equation by converting it to an initial value problem.

$$u''(x) = -x + \int_0^x (x-t)u(t)dt, \quad u(0) = 0, \ u'(0) = 1. \tag{100}$$

Differentiating both sides of (100) and using Leibniz rule we find

$$u'''(x) = -1 + \int_0^x u(t)\,dt, \tag{101}$$

and by differentiating again to reduce the equation to a pure differential equation we obtain

$$u^{(iv)}(x) = u(x). \tag{102}$$

Combining the given initial conditions in (100) with the other initial conditions, obtained by substituting $x = 0$ in (100) and (101), we write

$$u(0) = 0, \ u'(0) = 1, \ u''(0) = 0, \ u'''(0) = -1. \tag{103}$$

The characteristic equation of (102) is

$$r^4 - 1 = 0, \tag{104}$$

which gives the roots

$$r = \pm 1, \pm i. \tag{105}$$

so that the general solution is given by

$$u(x) = A \cosh x + B \sinh x + C \cos x + D \sin x, \tag{106}$$

where A, B, C and D are constants to be determined. Using the initial conditions (103) to determine the numerical values for the constants A, B, C and D, we find the solution given by

$$u(x) = \sin x. \tag{107}$$

Example 3. Solve the following Volterra integro-differential equation by reducing the equation to an initial value problem.

$$u'(x) = 2 - \frac{1}{4}x^2 + \frac{1}{4}\int_0^x u(t)dt \quad u(0) = 0. \tag{108}$$

Differentiating both sides of (108) with respect to x and using Leibniz rule
to differentiate the integral at the right hand side we obtain

$$u''(x) = -\frac{1}{2}x + \frac{1}{4}u(x), \tag{109}$$

or equivalently

$$u''(x) - \frac{1}{4}u(x) = -\frac{1}{2}x, \tag{110}$$

with initial conditions given by

$$u(0) = 0, \quad u'(0) = 2, \tag{111}$$

where the second initial condition was obtained by substituting $x = 0$
in both sides of (108). The characteristic equation for the corresponding
homogeneous equation of (110) is

$$r^2 - \frac{1}{4} = 0, \tag{112}$$

which gives the roots

$$r = \pm\frac{1}{2} \tag{113}$$

so that the complementary solution is given by

$$u_c(x) = A\cos(\frac{x}{2}) + B\sin(\frac{x}{2}), \tag{114}$$

where A and B are constants to be determined. A particular solution $u_p(x)$
can be obtained by assuming that

$$u_p(x) = C + Dx. \tag{115}$$

Substituting (115) into (110) yields

$$C = 0, \quad D = 2. \tag{116}$$

Combining (114)-(116) yields

$$u(x) = A\cos(\frac{x}{2}) + B\sin(\frac{x}{2}) + 2x, \tag{117}$$

which gives

$$u(x) = 2x, \tag{118}$$

upon using the initial conditions (111).

Exercises 5.7

Solve the following Volterra integro-differential equations by converting the problem to *an initial value problem*

1. $u'(x) = e^x - \int_0^x u(t)\,dt, \quad u(0) = 1.$

2. $u'(x) = 1 - \int_0^x u(t)\,dt, \quad u(0) = 0.$

3. $u''(x) = -x - \dfrac{1}{2!}x^2 + \int_0^x (x - t)u(t)\,dt, \quad u(0) = 1, u'(0) = 1.$

4. $u''(x) = 1 - \dfrac{1}{2!}x^2 + \int_0^x (x - t)u(t)\,dt, \quad u(0) = 2, u'(0) = 0.$

5. $u''(x) = -\dfrac{1}{2!}x^2 - \dfrac{2}{3}x^3 + \int_0^x (x - t)u(t)\,dt, \quad u(0) = 1, u'(0) = 4.$

6. $u''(x) = -x - \dfrac{1}{8}x^2 + \int_0^x (x - t)u(t)\,dt, \quad u(0) = \dfrac{1}{4}, u'(0) = 1.$

7. $u'(x) = 1 + \sin x + \int_0^x u(t)\,dt, \quad u(0) = -1.$

5.8 Volterra Integro-Differential Equations of the First Kind

The standard form of the Volterra integro-differential equation of the first kind was introduced by Linz [25, 26]. The standard form of this integro-differential equation reads

$$\int_0^x K_1(x,t)u(t)dt + \int_0^x K_2(x,t)u^{(n)}(t)\,dt = f(x), K_2(x,t) \neq 0, \quad (119)$$

with given initial conditions. The Volterra integro-differential equation of the first kind (119) was investigated by Linz [25, 26] by using analytical and numerical methods. Moreover, the equation was handled by other techniques as will be presented later.

The most significant method for handling this problem is to convert it to an equivalent Volterra integro-differential equation of the second kind, or even Volterra integral equation of the second kind simply by using the Leibniz rule for differentiating any integral. Also this can be achieved by integrating the second integral in (119) by parts. Having converted the Volterra integro-differential equation of the first kind to its equivalent of the second kind, then we can employ any of the methods that were applied earlier in this text, such as the series method, the Adomian decomposition method and the variational iteration method. For practical use, we will apply the latter method. The reader is advised to use other methods.

In this text we will concern ourselves on the Volterra integro-differential equation of the first kind where the kernels $K_1(x,t)$ and $K_2(x,t)$ of (119) are *difference kernels*, that each depends on the difference $(x-t)$ such as $(x-t+1), \sin(x-t), e^{x-t}, \ldots$ One more essential condition that should be considered that the solution exists if $K_2(x,x) \neq 0$. This necessary condition can be observed by introducing the conversion process to an equivalent Volterra integro-differential equation of the second kind.

Differentiating both sides of (119), and using Leibniz rule we find

$$
\begin{aligned}
u^{(n)}(x) &= \frac{f'(x)}{K_2(x,x)} - \frac{K_1(x,x)}{K_2(x,x)} u(x) - \frac{1}{K_2(x,x)} \int_0^x \frac{\partial(K_1(x,t))}{\partial x} u^{(n)}(t) dt \\
&- \frac{1}{K_2(x,x)} \int_0^x \frac{\partial(K_2(x,t))}{\partial x} u^{(n)}(t) \, dt.
\end{aligned}
$$

(120)

The obtained equation is a Volterra integro-differential equation of the second kind that can be evaluated only if $K_2(x,x) \neq 0$.

We will focus our illustrative examples by the variational iteration method, where other methods can be used as well. Recall that we need the Lagrange multiplier λ that will be used in the correction functional. Moreover, the zeroth approximation can be selected by using the given initial condition as presented before. The Volterra integro-differential equation of the first kind will be examined by using the variational iteration method when studying the following examples.

Example 1. Solve the Volterra integro-differential equation of the first kind

$$
\int_0^x (x - t + 1) u'(t) \, dt = 2e^x - x - 2, u(0) = 1.
$$

(121)

Differentiating both sides of this equation once with respect to x gives the Volterra integro-differential equation of the second kind

$$
u'(x) = 2e^x - 1 - \int_0^x u'(t) \, dt.
$$

(122)

The correction functional for equation (122) is given by

$$
u_{n+1}(x) = u_n(x) - \int_0^x \left(u_n'(t) - 2e^t + 1 + \int_0^t u_n'(r) \, dr \right) dt
$$

(123)

where we used $\lambda(t) = -1$. The zeroth approximation $u_0(x)$ can be selected by $u_0(x) = 1$. This gives the successive approximations

$$
\begin{aligned}
u_0(x) &= 1, \\
u_1(x) &= 2e^x - 1 - x, \\
u_2(x) &= 1 + x + \tfrac{1}{2!}x^2, \\
u_3(x) &= 1 + x + \tfrac{1}{2!}x^2 + \tfrac{1}{3!}x^3 + \tfrac{1}{4!}x^4,
\end{aligned}
\tag{124}
$$

$$\vdots$$

This gives

$$
u_n(x) = 1 + x + \frac{1}{2!}x^2 + \frac{1}{3!}x^3 + \frac{1}{4!}x^4 + \cdots,
\tag{125}
$$

that converges to the exact solution

$$
u(x) = e^x.
\tag{126}
$$

Example 2. Solve the Volterra integro-differential equation of the first kind

$$
\int_0^x (x - t)u(t)dt + \int_0^x (x - t + 1)u'(t)\,dt = 1 + x - \cos x, \, u(0) = 0.
\tag{127}
$$

Differentiating both sides of this equation once with respect to x gives the Volterra integro-differential equation of the second kind

$$
u'(x) = 1 + \sin x - \int_0^x (u(t) + u'(t))\,dt.
\tag{128}
$$

The correction functional for equation this equation is given by

$$
u_{n+1}(x) = u_n(x) - \int_0^x \left(u'_n(t) - 1 - \sin t + \int_0^t (u_n(r) + u'_n(r))\,dr \right) dt.
\tag{129}
$$

The zeroth approximation $u_0(x)$ can be selected by $u_0(x) = 0$. This gives the successive approximations

$$
\begin{aligned}
u_0(x) &= 0, \\
u_1(x) &= 1 + x - \cos x, \\
u_2(x) &= x - \tfrac{1}{3!}x^3 - \tfrac{1}{12}x^4 + \tfrac{1}{5!}x^5 + \cdots, \\
u_3(x) &= x - \tfrac{1}{3!}x^3 + \tfrac{1}{5!}x^5 - \tfrac{1}{7!}x^7 + \cdots,
\end{aligned}
\tag{130}
$$

$$\vdots$$

This gives the exact solution

$$
u(x) = \sin x.
\tag{131}
$$

Example 3. Use the variational iteration method to solve the Volterra integro-differential equation of the first kind

$$\int_0^x (x-t)u(t)dt - \frac{1}{2}\int_0^x (x-t+1)u'(t)\, dt = \frac{5}{2}x^2 - \frac{1}{2}x, \quad u(0) = 6. \quad (132)$$

Differentiating both sides of (132) once with respect to x gives the Volterra integro-differential equation of the second kind

$$u'(x) = -10x + 1 + \int_0^x (2u(t) - u'(t))\, dt. \quad (133)$$

The correction functional for equation this equation is given by

$$u_{n+1}(x) = u_n(x) - \int_0^x \left(u_n'(t) + 10t - 1 - \int_0^t (2u_n(r) - u_n'(r))\, dr \right)\, dt \quad (134)$$

where we used $\lambda(t) = -1$. The zeroth approximation $u_0(x)$ can be selected by $u_0(x) = 6$. This gives the successive approximations

$$
\begin{aligned}
u_0(x) &= 6, \\
u_1(x) &= 6 + x + x^2, \\
u_2(x) &= 6 + x + \frac{1}{2!}x^2 + \frac{1}{3!}x^4, \\
u_3(x) &= 6 + x + \frac{1}{2!}x^2 + \frac{1}{3!}x^3 + \frac{1}{12}x^4 - \frac{1}{30}x^5 + \cdots, \\
u_4(x) &= 6 + x + \frac{1}{2!}x^2 + \frac{1}{3!}x^3 + \frac{1}{4!}x^4 + \frac{1}{90}x^6 + \cdots, \\
u_5(x) &= 6 + x + \frac{1}{2!}x^2 + \frac{1}{3!}x^3 + \frac{1}{4!}x^4 + \frac{1}{5!}x^5 + \cdots, \\
&\quad \vdots
\end{aligned}
\quad (135)
$$

This gives

$$u_n(x) = 5 + \left(1 + x + \frac{1}{2!}x^2 + \frac{1}{3!}x^3 + \frac{1}{4!}x^4 + \frac{1}{5!}x^5 + \cdots \right). \quad (136)$$

The exact solution is therefore given by

$$u(x) = 5 + e^x. \quad (137)$$

Example 4. Solve the Volterra integro-differential equation of the first kind

$$\int_0^x (x-t+1)\, u'(t)\, dt = 1 + 2x + \frac{1}{2}x^2 - e^x, \quad u(0) = -1. \quad (138)$$

Differentiating both sides of (138) once with respect to x gives the Volterra integro-differential equation of the second kind

$$u'(x) = 2 + x - e^x - \int_0^x u'(t)\, dt. \quad (139)$$

The correction functional for equation (139) is given by

$$u_{n+1}(x) = u_n(x) - \int_0^x \left(u_n'(t) + e^t - t - 2 + \int_0^t u_n'(r)\, dr \right) dt \qquad (140)$$

where we used $\lambda(t) = -1$. The zeroth approximation $u_0(x)$ can be selected by $u_0(x) = -1$. This gives the successive approximations

$$
\begin{aligned}
u_0(x) &= -1, \\
u_1(x) &= 2x - \tfrac{1}{2}x^2 - e^x, \\
u_2(x) &= -1 + x - \tfrac{1}{2!}x^2 - \tfrac{1}{4!}x^4, \\
u_3(x) &= -1 + x - \tfrac{1}{2!}x^2 - \tfrac{1}{4!}x^4 - \tfrac{1}{6!}x^6 + \cdots, \\
u_4(x) &= -1 + x - \tfrac{1}{2!}x^2 - \tfrac{1}{4!}x^4 - \tfrac{1}{6!}x^6 - \tfrac{1}{8!}x^8 + \cdots,
\end{aligned}
\qquad (141)
$$

$$\vdots$$

This gives

$$u_n(x) = x - \left(1 + \frac{1}{2!}x^2 + \frac{1}{4!}x^4 + \frac{1}{6!}x^6 + \frac{1}{8!}x^8 + \cdots \right). \qquad (142)$$

The exact solution is therefore given by

$$u(x) = x - \cosh x. \qquad (143)$$

Exercises 5.8

Solve the following Volterra integro-differential equations of the first kind by using *the variational iteration method or any other method*

1. $\displaystyle\int_0^x (x - t + 1)u'(t)\, dt = \sin x + \cos x - 1 - x,\ u(0) = 1$

2. $\displaystyle\int_0^x (x - t + 1)u'(t)\, dt = -x,\ u(0) = 1$

3. $\displaystyle\int_0^x (x - t + 1)u'(t)\, dt = \sinh x + \cosh x - 1,\ u(0) = 0$

4. $\displaystyle\int_0^x (x - t)u(t)\, dt + \int_0^x (x - t + 1)u'(t)\, dt = 3e^x - 3 - 2x,\ u(0) = 1$

5. $\displaystyle\int_0^x (x - t)u(t)\, dt + \int_0^x (x - t + 1)u'(t)\, dt = \sin x + \frac{1}{2}x^2 + \frac{1}{6}x^3,\ u(0) = 1$

6. $\displaystyle\int_0^x (x-t)u(t)\, dt + \int_0^x (x-t+2)u'(t)\, dt = \sinh x + 3\cosh x - x - 3,\ u(0) = 1$

Chapter 6

Singular Integral Equations

6.1 Introduction

An integral equation is called a *singular* integral equation if one or both limits of integration become *infinite*, or if the kernel $K(x,t)$ of the equation becomes *infinite* at one or more points in the interval of integration. In other words, the integral equation of the first kind

$$f(x) = \lambda \int_{\alpha(x)}^{\beta(x)} K(x,t)\, u(t) dt \qquad (1)$$

or the integral equation of the second kind

$$u(x) = f(x) + \lambda \int_{\alpha(x)}^{\beta(x)} K(x,t)\, u(t) dt, \qquad (2)$$

is called *singular* if the lower limit $\alpha(x)$, the upper limit $\beta(x)$ or both limits of integration are *infinite*. Moreover, the equation (1) or (2) is also called a *singular* integral equation if the kernel $K(x,t)$ becomes *infinite* at one or more points in the domain of integration. Examples of the first style of *singular* integral equations are given by the following examples:

$$u(x) = 1 + e^{-x} - \int_0^\infty u(t) dt, \qquad (3)$$

$$F(\lambda) = \int_{-\infty}^\infty e^{-\imath \lambda x} u(x) dx, \qquad (4)$$

171

$$L[u(x)] \;=\; \int_0^\infty e^{-\lambda x} u(x) dx. \tag{5}$$

The integral equations (4) and (5) are Fourier transform and Laplace transform of the function $u(x)$ respectively. In addition, the equations (4) and (5) are in fact Fredholm integral equations of the first kind with kernels given by $K(x,t) = e^{-i\lambda x}$ and $K(x,t) = e^{-\lambda x}$. It is important to note that the Laplace transforms and the Fourier transforms are usually used for solving ordinary and partial differential equations with constant coefficients. However, these transforms will not be used in this text to solve integral equations, but will be used in the derivation of two formulas as will be seen later.

One important point to be noted here is that the singular behavior in (3)-(5) has been attributed to the range of integration becoming *infinite*.

Examples of the second style of *singular* integral equations are given by

$$x^2 = \int_0^x \frac{1}{\sqrt{x-t}} u(t) dt, \tag{6}$$

$$x = \int_0^x \frac{1}{(x-t)^\alpha} u(t) dt, \quad 0 < \alpha < 1, \tag{7}$$

$$u(x) = 1 + 2\sqrt{x} - \int_0^x \frac{1}{\sqrt{x-t}} u(t) dt, \tag{8}$$

where the singular behavior in this style of equations has been attributed to the kernel $K(x,t)$ becoming *infinite* as $t \to x$.

It is important to note that integral equations similar to examples (6) and (7) are called Abel's problems and generalized Abel's integral equations respectively. Moreover these styles of singular integral equations are among the earliest integral equations established by the Norwegian mathematician Niels Abel in 1823. In addition, Abel's equations arise frequently in mathematical physics.

However, singular equations similar to example (8) are called the weakly-singular second-kind Volterra-type integral equations. This type of equations usually arise in scientific and engineering applications like heat conduction, super fluidity and crystal growth. Recently, the weakly-singular second-kind Volterra type integral equations have been the subject of extensive analytical studies. Moreover, numerical studies have been carried out to obtain approximations, of high accuracy level ,to the exact solution.

In addition to the definitions of the singular Volterra integral equations that we defined, we note that singularity behavior arises also in Fredholm integral equations of the second kind. Examples of the weakly-singular Fredholm integral equations are given by

$$u(x) = f(x) + \int_a^b \frac{1}{\sqrt{x-t}}\, u(t)\, dt, \qquad (9)$$

and

$$u(x) = f(x) + \int_a^b \frac{1}{(x-t)^\alpha}\, u(t)\, dt, 0 < \alpha < 1. \qquad (10)$$

Recall that for the first kind, the unknown solution $u(x)$ appears only inside the integral sign, whereas for the second kind the unknown solution $u(x)$ appears inside and outside the integral sign, a characteristic feature of the first and the second kind integral equations.

We point out that the weakly singular Volterra integral equations arise in mathematical physics applications, chemical applications such as stereology, heat conduction, radiation of heat from a semi-infinite solid and crystal growth. However, the weakly-singular Fredholm integral equations often arise in scientific applications such as Dirichlet problems, radiation equilibrium applications, electrostatics, potential theory, astrophysics, and radiative heat transfer.

In this chapter we will focus our study on the second style of singular Volterra integral equations, namely the equations where the kernel $K(x,t)$ becomes unbounded at one or more points of singularities in its domain of definition. The equations that will be investigated are Abel's problem, generalized Abel integral equations and the weakly-singular second-kind Volterra type integral equations. Moreover, we will proceed in our study to cover the singular Fredholm integral equations using the practical methods. The singular Fredholm integral equations was approached by many approaches, mostly by numerical techniques, such as homotopy method, wavelet Galerkin method, Taylor expansion method and other methods as well. In a manner parallel to the approach used in previous chapters, we will focus our study on the techniques that will guarantee the existence of a unique solution to any singular integral equation with singularity related to the kernel $K(x,t)$ becoming unbounded at its domain of integration. We point out here that singular integral equations are in general very difficult to handle.

6.2 Abel's Problem

Abel in 1823 investigated the motion of a particle that slides down along a smooth unknown curve, in a vertical plane, under the influence of the gravitational field. It is assumed that the particle starts from rest at a point P, with vertical elevation x, slides along the unknown curve, to the lowest point O on the curve where the vertical distance is $x = 0$. The total time of descent T from the highest point to the lowest point on the curve is given in advance, and dependent on the elevation x, hence expressed by

$$T = h(x). \tag{11}$$

Assuming that the curve between the points P and O has an arclength s, then the velocity at a point Q on the curve, between P and O, given by

$$\frac{ds}{dT} = -\sqrt{2g(x - t)}, \tag{12}$$

where t is a variable coordinate defines the vertical distance of the point Q, and g is a constant defines the acceleration of gravity. Integrating both sides of (12) gives

$$T = -\int_O^P \frac{ds}{\sqrt{2g(x - t)}}. \tag{13}$$

Setting

$$ds = u(t)dt, \tag{14}$$

and using (11) we find that the equation of motion of the sliding particle is governed by

$$f(x) = \int_0^x \frac{1}{\sqrt{x - t}} u(t)dt. \tag{15}$$

We point out that $f(x)$ is a predetermined function that depends on the elevation x and given by

$$f(x) = \sqrt{2g}\, h(x), \tag{16}$$

where g is the gravitational constant, and $h(x)$ is the time of descent from the highest point to the lowest point on the curve. The main goal of Abel's problem is to determine the unknown function $u(x)$ under the integral sign that will define the equation of the curve. Having determined $u(x)$, the equation of the smooth curve, where the particle slides along, can be easily obtained using the calculus formulas related to the arclength concepts.

It is worth mentioning that Abel's integral equation (15) is also called Volterra integral equation of the first kind. Besides, the kernel $K(x, t)$ in (15) is given by

$$K(x, t) = \frac{1}{\sqrt{x - t}}, \tag{17}$$

which shows that the kernel (17) is singular in that

$$K(x, t) \to \infty \quad \text{as} \quad t \to x. \tag{18}$$

The interesting Abel's problem has been approached by different methods. In the following we will employ Laplace transforms only to determine a suitable formula to solve Abel's problem (15), noting that Laplace transforms will not be used in our approach to handle the singular equations. Taking Laplace transforms of both sides of (15) leads to

$$
\begin{aligned}
L[f(x)] &= L[u(x)] \, L[x^{-\frac{1}{2}}] \\
&= L[u(x)] \frac{\Gamma(1/2)}{z^{1/2}},
\end{aligned} \tag{19}
$$

where Γ is the gamma function. In Appendix D, the definition of the gamma function and some of the relations related to it are given. Noting that $\Gamma(\frac{1}{2}) = \sqrt{\pi}$, the equation (19) becomes

$$
L[u(x)] = \frac{z^{\frac{1}{2}}}{\sqrt{\pi}} L[f(x)], \tag{20}
$$

which can be rewritten by

$$
L[u(x)] = \frac{z}{\pi} \left(\sqrt{\pi} z^{-\frac{1}{2}} L[f(x)] \right). \tag{21}
$$

Setting

$$
h(x) = \int_0^x (x - t)^{-\frac{1}{2}} f(t) dt, \tag{22}
$$

into (21) yields

$$
L[u(x)] = \frac{z}{\pi} L[h(x)], \tag{23}
$$

which gives

$$
L[u(x)] = \frac{1}{\pi} L[h'(x)], \tag{24}
$$

upon using the fact

$$
L[h'(x)] = z \, L[h(x)]. \tag{25}
$$

Applying L^{-1} to both sides of (24) yields the easily calculable formula

$$
u(x) = \frac{1}{\pi} \frac{d}{dx} \int_0^x \frac{f(t)}{\sqrt{x - t}} dt, \tag{26}
$$

that will be used for the determination of the solution. It is clear that Leibniz rule is not applicable in (26) because the integrand is discontinuous at the interval of integration. As indicated earlier, determination of $u(x)$ will lead to the determination of the curve where the particle slides along this curve.

It is obvious that Abel's problem given by (15) can be solved now by using the formula (26) where the unknown function $u(x)$ has been replaced by the given function $f(x)$. One last remark concerns the use of the formula (26). The process consists of selecting the proper substitution for $(x - t)$, integrate the resulting definite integral and finally differentiate the result of the evaluation. Appendix B, an appropriate calculator or any symbolic computer software, such as Maple or Mathematica, can be used as a helpful tool needed for evaluating the integrals involved.

The procedure of using the formula (26) that determines the solution of Abel's problem (15) will be illustrated by the following examples.

Example 1. As a first example we consider the following Abel's problem

$$\pi = \int_0^x \frac{1}{\sqrt{x - t}} u(t)\, dt. \tag{27}$$

Substituting $f(x) = \pi$ in (26) yields

$$
\begin{aligned}
u(x) &= \frac{1}{\pi} \frac{d}{dx} \int_0^x \frac{\pi}{\sqrt{x - t}}\, dt \\
&= \frac{d}{dx} \int_0^x \frac{1}{\sqrt{x - t}}\, dt.
\end{aligned}
\tag{28}
$$

Setting the substitution $y = x - t$ in (28), we obtain

$$
\begin{aligned}
u(x) &= \frac{d}{dx} \left(2\sqrt{x} \right) \\
&= \frac{1}{\sqrt{x}}.
\end{aligned}
\tag{29}
$$

Example 2. Solve the following Abel's problem

$$\frac{\pi}{2} x = \int_0^x \frac{1}{\sqrt{x - t}} u(t)\, dt. \tag{30}$$

Substituting $f(x) = \frac{\pi}{2} x$ in (26) gives

$$
\begin{aligned}
u(x) &= \frac{1}{\pi} \frac{d}{dx} \int_0^x \frac{\frac{\pi}{2} t}{\sqrt{x - t}}\, dt \\
&= \frac{1}{2} \frac{d}{dx} \int_0^x \frac{t}{\sqrt{x - t}}\, dt.
\end{aligned}
\tag{31}
$$

Using integration by substitution, where we set $y = x - t$, or by using Appendix B, we obtain

$$
\begin{aligned}
u(x) &= \frac{1}{2}\frac{d}{dx}\left(\frac{4}{3}x^{\frac{3}{2}}\right) \\
&= x^{\frac{1}{2}}.
\end{aligned}
\tag{32}
$$

Example 3. As a third example we consider the following Abel's problem

$$
2\sqrt{x} = \int_0^x \frac{1}{\sqrt{x-t}}u(t)\,dt.
\tag{33}
$$

Substituting $f(x) = 2\sqrt{x}$ in (26) we find

$$
u(x) = \frac{2}{\pi}\frac{d}{dx}\int_0^x \frac{\sqrt{t}}{\sqrt{x-t}}dt.
\tag{34}
$$

The integral at the right hand side of (34) can be evaluated by using integration by substitution, where in this case we set the substitution

$$
t = x\sin^2\theta,
\tag{35}
$$

so that

$$
\sqrt{x-t} = \sqrt{x}\cos\theta,
\tag{36}
$$

and

$$
dt = 2x\sin\theta\cos\theta d\theta.
\tag{37}
$$

Substituting (35)-(37) in (34) we obtain

$$
\begin{aligned}
u(x) &= \frac{4}{\pi}\frac{d}{dx}\left(x\int_0^{\pi/2}\sin^2\theta d\theta\right) \\
&= \frac{4}{\pi}\frac{d}{dx}\left(x\left[\frac{1}{2}\theta - \frac{1}{4}\sin(2\theta)\right]_0^{\pi/2}\right) \\
&= 1.
\end{aligned}
\tag{38}
$$

The integral at the right hand side of (38) may be evaluated directly by using Appendix B.

Exercises 6.2

Solve the following Abel's integral equations:

1. $\quad \pi(x+1) = \int_0^x \frac{1}{\sqrt{x-t}} u(t)\, dt.$

2. $\quad \dfrac{\pi}{2}(x^2 - x) = \int_0^x \frac{1}{\sqrt{x-t}} u(t)\, dt.$

3. $\quad x^2 + x + 1 = \int_0^x \frac{1}{\sqrt{x-t}} u(t)\, dt.$

4. $\quad \dfrac{3\pi}{8} x^2 = \int_0^x \frac{1}{\sqrt{x-t}} u(t)\, dt.$

5. $\quad \dfrac{4}{3} x^{\frac{3}{2}} = \int_0^x \frac{1}{\sqrt{x-t}} u(t)\, dt.$

6. $\quad \dfrac{8}{15} x^{\frac{5}{2}} = \int_0^x \frac{1}{\sqrt{x-t}} u(t)\, dt.$

7. $\quad x^3 = \int_0^x \frac{1}{\sqrt{x-t}} u(t)\, dt.$

8. $\quad x^4 = \int_0^x \frac{1}{\sqrt{x-t}} u(t)\, dt.$

9. $\quad x + x^3 = \int_0^x \frac{1}{\sqrt{x-t}} u(t)\, dt.$

10. $\quad \sin x = \int_0^x \frac{1}{\sqrt{x-t}} u(t)\, dt.$

6.3   The Generalized Abel's Integral Equation

It is important here to note that Abel introduced the more general singular integral equation

$$f(x) = \int_0^x \frac{1}{(x-t)^\alpha} u(t)\, dt, \quad 0 < \alpha < 1, \qquad (39)$$

known as the *Generalized Abel's integral equation*, where the exponent of the denominator of the kernel is α, such as $0 < \alpha < 1$. It can be easily seen that Abel's problem discussed above is a special case of the *generalized equation* where $\alpha = \frac{1}{2}$. To determine a practical formula for the solution $u(x)$ of (39), and hence for the Abel's problem, we simply use the Laplace transform in a similar manner to that used above. As noted before, the Laplace transform will be used for the derivation of the proper formula, but will not be used in handling the equations. Taking Laplace transforms to both sides (39) yields

$$
\begin{aligned}
L[f(x)] \;&=\; L[u(x)]\, L[x^{-\alpha}] \\[2mm]
&=\; L[u(x)] \frac{\Gamma(1-\alpha)}{z^{1-\alpha}},
\end{aligned}
\qquad (40)
$$

where Γ is the gamma function. The equation (40) can be written as

$$L[u(x)] = \frac{z}{\Gamma(\alpha)\Gamma(1-\alpha)}\Gamma(\alpha)z^{-\alpha}L[f(x)], \qquad (41)$$

or equivalently

$$L[u(x)] = \frac{z}{\Gamma(\alpha)\Gamma(1-\alpha)}L[g(x)], \qquad (42)$$

where

$$g(x) = \int_0^x (x-t)^{\alpha-1}f(t)dt. \qquad (43)$$

Using (43) into (42) yields

$$L[u(x)] = \frac{\sin(\alpha\pi)}{\pi}L[g^{'}(x)], \qquad (44)$$

upon using the identities

$$L[g^{'}(x)] = z\,L[g(x)], \qquad (45)$$

and

$$\Gamma(\alpha)\Gamma(1-\alpha) = \frac{\pi}{\sin(\alpha\pi)}, \qquad (46)$$

from Laplace transforms and Appendix D respectively. Applying L^{-1} to both sides of (44) yields the easily computable formula for determining the solution

$$u(x) = \frac{\sin(\alpha\pi)}{\pi}\frac{d}{dx}\int_0^x \frac{f(t)}{(x-t)^{1-\alpha}}dt, \quad 0 < \alpha < 1. \qquad (47)$$

Recall that $f(x)$ is differentiable,therefore we can derive from (47) a more suitable formula that will support our computational purposes. To determine this formula, we first integrate the integral at the right hand side of (47) by parts where we obtain

$$\int_0^x \frac{f(t)}{(x-t)^{1-\alpha}}dt = -\frac{1}{\alpha}[f(t)(x-t)^\alpha]_0^x + \frac{1}{\alpha}\int_0^x (x-t)^\alpha f^{'}(t)dt$$
$$= \frac{1}{\alpha}f(0)x^\alpha + \frac{1}{\alpha}\int_0^x (x-t)^\alpha f^{'}(t)dt. \qquad (48)$$

Differentiating both sides of (48), noting that Leibniz rule should be used in differentiating the integral at the right hand side, yields

$$\frac{d}{dx}\int_0^x \frac{f(t)}{(x-t)^{1-\alpha}}dt = \frac{f(0)}{x^{1-\alpha}} + \int_0^x \frac{f^{'}(t)}{(x-t)^{1-\alpha}}dt. \qquad (49)$$

Substituting (49) into (47) yields the desired formula given by

$$u(x) = \frac{\sin(\alpha\pi)}{\pi}\left(\frac{f(0)}{x^{1-\alpha}} + \int_0^x \frac{f'(t)}{(x-t)^{1-\alpha}}dt\right), \quad 0 < \alpha < 1, \qquad (50)$$

that will be used to determine the solution of the generalized Abel's equation and consequently, of the standard Abel's problem as well. This will be illustrated by examining the following examples.

Example 1. Solve the following generalized Abel's integral equation

$$27x^{\frac{8}{3}} = \int_0^x \frac{1}{(x-t)^{\frac{1}{3}}}u(t)\,dt. \qquad (51)$$

Notice that $\alpha = \frac{1}{3}, f(x) = 27x^{\frac{8}{3}}$. Using (47) gives

$$u(x) = \frac{\sqrt{3}}{2\pi}\frac{d}{dx}\int_0^x \frac{27t^{\frac{8}{3}}}{(x-t)^{\frac{2}{3}}}dt = 40x^2. \qquad (52)$$

Example 2. Solve the following generalized Abel's integral equation

$$4\sqrt{3}\pi x = \int_0^x \frac{1}{(x-t)^{\frac{2}{3}}}u(t)\,dt. \qquad (53)$$

Notice that $\alpha = \frac{2}{3}, f(x) = 4\sqrt{3}\pi x$. Using (47) gives

$$u(x) = \frac{\sqrt{3}}{2\pi}\frac{d}{dx}\int_0^x \frac{4\sqrt{3}\pi t}{(x-t)^{\frac{1}{3}}}dt = 9x^{\frac{2}{3}}. \qquad (54)$$

Example 3. Solve the following generalized Abel's integral equation

$$\frac{9}{10}x^{\frac{5}{3}} = \int_0^x \frac{1}{(x-t)^{\frac{1}{3}}}u(t)\,dt. \qquad (55)$$

Notice that $\alpha = \frac{1}{3}, f(x) = \frac{9}{10}x^{\frac{5}{3}}$. Using (47) gives

$$u(x) = \frac{\sqrt{3}}{2\pi}\frac{d}{dx}\int_0^x \frac{4\frac{9}{10}t^{\frac{5}{3}}}{(x-t)^{\frac{2}{3}}}dt = x. \qquad (56)$$

Example 4. Find an approximate solution to the following Abel's problem

$$\sinh x = \int_0^x \frac{1}{\sqrt{x-t}}u(t)\,dt. \qquad (57)$$

In this example $f(x) = \sinh x$, hence $f(0) = 0$ and $f'(x) = \cosh x$. Using the formula (47) we find

$$u(x) = \frac{1}{\pi} \int_0^x \frac{\cosh t}{\sqrt{x-t}} dt. \tag{58}$$

An approximate solution can be found by considering $\cosh x \approx 1 + \frac{x^2}{2!}$ for small x. Consequently, we have

$$u(x) \approx \frac{1}{\pi} \int_0^x \frac{1 + \frac{t^2}{2}}{\sqrt{x-t}} dt, \tag{59}$$

which gives

$$u(x) \approx \frac{2}{15\pi} \sqrt{x}(15 + 4x^2), \quad \text{for small } x, \tag{60}$$

by integrating by substitution or by using Appendix B.

Exercises 6.3

Solve the following generalized Abel's integral equations

1. $\dfrac{36}{55} x^{\frac{11}{6}} = \displaystyle\int_0^x \frac{1}{(x-t)^{\frac{1}{6}}} u(t)\, dt$

2. $\dfrac{243}{440} x^{\frac{11}{3}} = \displaystyle\int_0^x \frac{1}{(x-t)^{\frac{1}{3}}} u(t)\, dt$

3. $24x^{\frac{1}{4}} = \displaystyle\int_0^x \frac{1}{(x-t)^{\frac{3}{4}}} u(t)\, dt$

4. $3\pi x^{\frac{1}{3}} + 9x^{\frac{4}{3}} = \displaystyle\int_0^x \frac{1}{(x-t)^{\frac{2}{3}}} u(t)\, dt$

5. $\dfrac{36}{7} x^{\frac{7}{6}} = \displaystyle\int_0^x \frac{1}{(x-t)^{\frac{5}{6}}} u(t)\, dt$

6. $27x^{\frac{7}{3}} + 9x^{\frac{4}{3}} = \displaystyle\int_0^x \frac{1}{(x-t)^{\frac{2}{3}}} u(t)\, dt$

6.4 The Weakly-Singular Volterra Integral Equations

As indicated earlier, the weakly-singular Volterra integral equations of the second kind are given by

$$u(x) = g(x) + \int_0^x \frac{\beta}{\sqrt{x-t}} u(t) dt, \quad x \in [0, T], \tag{61}$$

and the generalized form

$$u(x) = g(x) + \int_0^x \frac{\beta}{(x-t)^\alpha} u(t)dt, \quad x \in [0,T], 0 < \alpha < 1, \qquad (62)$$

appear frequently in many mathematical physics and chemistry applications such as heat conduction, crystal growth, electro chemistry, and radiation of heat from a semi-infinite solid. It is to be noted that β is a constant and $T = 1, 2$, or 3 depending on the science model under discussion. It is also assumed that the function $g(x)$ is sufficiently smooth so that a unique solution to (61) is guaranteed. The equations (61) and (62) fall under the category of singular equations with singular kernel $K(x,t) = \frac{1}{\sqrt{x-t}}$. Notice that the kernel is called weakly singular as the singularity may be transformed away by a change of variables [12]. A considerable amount of work has been carried out recently on these models to determine its exact solutions or to achieve numerical approximations of high degree of accuracy.

In this section we will base our discussion on the Adomian decomposition method that was introduced in the preceding chapters, and mostly we will use the modified decomposition method and the noise terms phenomenon. We will show that this technique is an effective and powerful tool to handle this style of singular equations analytically and numerically. The method has been discussed extensively and need not be introduced in details here. We note here that we cannot use the variational iteration method because we cannot use Leibniz rule for singular equations.

6.4.1   The Adomian Decomposition Method

In the following we outline a brief framework of the method. The method as presented before gives the solution in a series that converges to the exact solution if an exact solution exists. To determine the solution $u(x)$ of (61) we usually use the decomposition

$$u(x) = \sum_{n=0}^{\infty} u_n(x), \qquad (63)$$

into both sides of (61) to obtain

$$\sum_{n=0}^{\infty} u_n(x) = g(x) + \int_0^x \frac{\beta}{\sqrt{x-t}} \left(\sum_{n=0}^{\infty} u_n(t) \right) dt, \quad x \in [0,T]. \qquad (64)$$

The components $u_0(x), u_1(x), u_2(x), \dots$ are immediately determined upon applying the following recurrent algorithm

$$
\begin{cases}
u_0(x) &= g(x), \\[2mm]
u_1(x) &= \displaystyle\int_0^x \frac{\beta}{\sqrt{x-t}} u_0(t)dt, \\[2mm]
u_2(x) &= \displaystyle\int_0^x \frac{\beta}{\sqrt{x-t}} u_1(t)dt, \\[2mm]
&\quad\vdots
\end{cases}
\tag{65}
$$

The same approach can be used for the generalized weakly-singular integral equation where $0 < \alpha < 0$. Having determined the components $u_i(x), i \geq 0$, the solution $u(x)$ of (61) will be easily obtained in the form of a rapid convergent power series by substituting the derived components in (63).

It is important to note that the phenomenon of the self-cancelling noise terms, where like terms with opposite signs appear in specific problems, should be observed here between the components $u_0(x)$ and $u_1(x)$. As mentioned earlier, the appearance of these terms usually speeds the convergence of the solution and normally minimizes the size of the computational work. For illustration purposes, we discuss the following examples.

Example 1. We first consider the weakly-singular Volterra integral equation of the second kind

$$
u(x) = \sqrt{x} + \frac{1}{2}\pi x - \int_0^x \frac{1}{\sqrt{x-t}} u(t)\, dt, \quad I = [0, 2].
\tag{66}
$$

Using the recurrent algorithm we set

$$
\begin{aligned}
u_0(x) &= \sqrt{x} + \frac{1}{2}\pi x, \\[2mm]
u_1(x) &= -\int_0^x \frac{\sqrt{t} + \frac{1}{2}\pi t}{\sqrt{x-t}}\, dt \\[2mm]
&= -\frac{1}{2}\pi x - \frac{2}{3}\pi x^{3/2}.
\end{aligned}
\tag{67}
$$

The result (67) can be obtained directly by using Appendix B, calculator, or computer software. Observing the appearance of the noise terms $\pm\frac{1}{2}\pi x$ between the components $u_0(x)$ and $u_1(x)$, and verifying that the non-cancelled term in $u_0(x)$ justifies the equation (66) yields the exact solution given by

$$
u(x) = \sqrt{x}.
\tag{68}
$$

It can be shown that it is possible to obtain the exact solution (68) by using the modified decomposition method. This can be done by splitting the nonhomogeneous part $g(x)$ into two parts. Accordingly, we set

$$u_0(x) = \sqrt{x}, \tag{69}$$

so that

$$u_1(x) = \frac{1}{2}\pi x - \int_0^x \frac{\sqrt{t}}{\sqrt{x-t}}\, dt, \tag{70}$$

which gives

$$u_1(x) = 0. \tag{71}$$

Consequently, other components will vanish, and the exact solution (68) follows immediately.

Example 2. As a second example we consider the weakly-singular second-kind Volterra integral equation

$$u(x) = x + \frac{4}{3}x^{3/2} - \int_0^x \frac{1}{\sqrt{x-t}}u(t)\, dt, \quad I = [0,2]. \tag{72}$$

Proceeding as before we set

$$
\begin{aligned}
u_0(x) &= x + \frac{4}{3}x^{3/2}, \\
u_1(x) &= -\int_0^x \frac{t + \frac{4}{3}t^{3/2}}{\sqrt{x-t}}\, dt \tag{73} \\
&= -\frac{4}{3}x^{3/2} - \frac{1}{2}\pi x^2.
\end{aligned}
$$

Cancelling the noise terms between the components $u_0(x)$ and $u_1(x)$, and verifying that the remaining term in $u_0(x)$ satisfies the equation (72) gives the exact solution

$$u(x) = x. \tag{74}$$

As discussed in example 1, we can obtain the exact solution (74) by using the modified decomposition method. We leave it as an exercise to the reader.

Example 3. We next consider the weakly-singular Volterra integral equation of the second kind

$$u(x) = 2\sqrt{x} - \int_0^x \frac{1}{\sqrt{x-t}}u(t)\, dt, \quad I = [0,2]. \tag{75}$$

Following the discussion in the previous examples we use the recurrence relation

$$
\begin{aligned}
u_0(x) &= 2\sqrt{x}, \\
u_1(x) &= -2\int_0^x \frac{\sqrt{t}}{\sqrt{x-t}}\,dt, \\
&= -\pi x, \\
u_2(x) &= \int_0^x \frac{\pi t}{\sqrt{x-t}}\,dt, \\
&= \frac{4}{3}\pi x^{3/2}, \\
u_3(x) &= -\frac{4}{3}\int_0^x \frac{\pi t^{3/2}}{\sqrt{x-t}}\,dt, \\
&= -\frac{1}{2}\pi^2 x^2,
\end{aligned}
\tag{76}
$$

$$\vdots$$

It is clear that the noise terms did not appear between the components $u_0(x)$ and $u_1(x)$. This explains why we determined more components. Combining the obtained results, the solution $u(x)$ in a series form

$$
u(x) = 2\sqrt{x} - \pi x + \frac{4}{3}\pi x^{3/2} - \frac{1}{2}\pi^2 x^2 + \cdots,
\tag{77}
$$

is readily obtained. The result obtained in (77) can be expressed as

$$
u(x) = 2\sqrt{x} - \pi x + O(x^{\frac{3}{2}}), \text{ as } x \to 0.
\tag{78}
$$

It is to be noted that in this example we determined four components of the series solution. Other components can be obtained in a similar fashion to increase the degree of accuracy for numerical purposes. However, the exact solution of (75) is given by

$$
u(x) = 1 - e^{\pi x}\operatorname{erfc}(\sqrt{\pi x}),
\tag{79}
$$

where $erfc$ is the complementary error function normally used in probability topics. The definitions of the error function and the complementary error function can be found in Appendix D.

Example 4. We now consider the weakly-singular Volterra integral equation of the second kind

$$
u(x) = x - \frac{9}{10}x^{\frac{5}{3}} + \int_0^x \frac{1}{(x-t)^{\frac{1}{3}}} u(t)\,dt.
\tag{80}
$$

In this example we will use the modified decomposition method. Hence, we use the recurrence relation

$$
\begin{aligned}
u_0(x) &= x, \\
u_1(x) &= -\frac{9}{10}x^{\frac{5}{3}} + \int_0^x \frac{t}{(x-t)^{\frac{1}{3}}} u(t)\,dt = 0.
\end{aligned}
\tag{81}
$$

This yields the exact solution

$$u(x) = x. \tag{82}$$

Example 5. We finally consider the weakly-singular Volterra integral equation of the second kind

$$u(x) = 1 + x^2 - \frac{4}{3}x^{\frac{3}{4}} - \frac{128}{231}x^{\frac{11}{4}} + \int_0^x \frac{1}{(x-t)^{\frac{1}{4}}} u(t)\, dt. \tag{83}$$

Using the modified decomposition method sets the recurrence relation

$$\begin{aligned}
u_0(x) &= 1 + x^2, \\
u_1(x) &= -\frac{4}{3}x^{\frac{3}{4}} - \frac{128}{231}x^{\frac{11}{4}} + \int_0^x \frac{(1+t^2)}{(x-t)^{\frac{1}{4}}} u(t)\, dt = 0.
\end{aligned} \tag{84}$$

This yields the exact solution

$$u(x) = 1 + x^2. \tag{85}$$

Exercises 6.4

Use the decomposition method or the modified decomposition method to solve the following weakly-singular Volterra integral equations of the second kind:

1. $u(x) = \sqrt{x} - \pi x + 2 \int_0^x \frac{1}{\sqrt{x-t}} u(t)\, dt.$

2. $u(x) = x^{\frac{3}{2}} + \frac{3}{8}\pi x^2 - \int_0^x \frac{1}{\sqrt{x-t}} u(t)\, dt.$

3. $u(x) = \frac{1}{2} - \sqrt{x} + \int_0^x \frac{1}{\sqrt{x-t}} u(t)\, dt.$

4. $u(x) = \sqrt{x} - \frac{1}{2}\pi x + \int_0^x \frac{1}{\sqrt{x-t}} u(t)\, dt.$

5. $u(x) = x^{\frac{5}{2}} - \frac{5}{16}\pi x^3 + \int_0^x \frac{1}{\sqrt{x-t}} u(t)\, dt.$

6. $u(x) = x^3 + \frac{32}{33}x^{\frac{7}{2}} - \int_0^x \frac{1}{\sqrt{x-t}} u(t)\, dt.$

7. $u(x) = 1 + x - 2\sqrt{x} - \frac{4}{3}x^{\frac{3}{2}} + \int_0^x \frac{1}{\sqrt{x-t}} u(t)\, dt.$

8. $u(x) = 1 + 2\sqrt{x} - \int_0^x \frac{1}{\sqrt{x-t}} u(t)\, dt.$

9. $u(x) = x^2 + \frac{16}{15}x^{\frac{5}{2}} - \int_0^x \frac{1}{\sqrt{x-t}} u(t)\, dt.$

10. $u(x) = \frac{2}{\pi}\sqrt{x} + \frac{15}{16}x^2 - x - x^{\frac{5}{2}} + \int_0^x \frac{1}{\sqrt{x-t}} u(t)\, dt.$

11. $\quad u(x) = 1 + x - \dfrac{4}{3}x^{\frac{3}{4}} - \dfrac{16}{21}x^{\frac{7}{4}} + \displaystyle\int_0^x \dfrac{1}{(x-t)^{\frac{1}{4}}}u(t)\,dt.$

12. $\quad u(x) = x + x^2 - \dfrac{9}{4}x^{\frac{4}{3}} - \dfrac{243}{140}x^{\frac{10}{3}} + \displaystyle\int_0^x \dfrac{1}{(x-t)^{\frac{2}{3}}}u(t)\,dt.$

13. $\quad u(x) = 1 + 3x^3 - \dfrac{4}{3}x^{\frac{3}{4}} - \dfrac{512}{385}x^{\frac{15}{4}} + \displaystyle\int_0^x \dfrac{1}{(x-t)^{\frac{1}{4}}}u(t)\,dt.$

14. $\quad u(x) = 5 - x - 6x^{\frac{5}{6}} + \dfrac{36}{55}x^{\frac{11}{6}} + \displaystyle\int_0^x \dfrac{1}{(x-t)^{\frac{1}{6}}}u(t)\,dt.$

6.5 The Weakly-Singular Fredholm Integral Equations

The weakly-singular Fredholm integral equations of the second kind take the form

$$u(x) = f(x) + \int_0^1 \frac{1}{\sqrt{x-t}}\,u(t)dt,\ x \in [0,1], \tag{86}$$

and can be generalized to

$$u(x) = f(x) + \int_0^1 \frac{1}{(x-t)^\alpha}\,u(t)dt, 0 < \alpha < 1,\ x \in [0,1], \tag{87}$$

where $u(t)$ is the unknown solution. The weakly-singular Fredholm integral equations often arise in scientific applications such as Dirichlet problems, radiation equilibrium applications, electrostatics, potential theory, astrophysics, and radiative heat transfer.

In the literature, there is a variety of numerical methods that were used for solving weakly-singular Fredholm integral equations (86)–(87), where the main goal was the determination of numerical approximations of the solutions. Examples of the methods that were used so far are the Galerkin method, collocation, quadrature methods, the homotopy analysis method, and the Taylor series method. The spectral method, the random point approximation method, and a fast spectral collocation method was applied for surface integral equations of potential problems in a spheroid. In [26], the spectral collocation method were also used for solving these equations.

In this chapter, we will use the modified decomposition method and the noise terms phenomenon to determine exact solutions of such singular integral equations. In the sequel, we will briefly review the necessary steps of the two proposed schemes that will be used.

6.5.1   The Modified Decomposition Method

The Adomian decomposition method admits the use of the decomposition
series for the solution

$$u(x) = \sum_{n=0}^{\infty} u_n(x),$$ (88)

where the solution components $u_i(x), i \geq 0$, are determined by using the
standard recurrence relation

$$
\begin{aligned}
u_0(x) &= f(x) \\
u_{n+1}(x) &= \int_a^b K(x,t)u_n(t)\, dt, n \geq 0.
\end{aligned}
$$ (89)

However, the modified decomposition method introduces a slight change in
the recurrence relation. The modified decomposition method decomposes
the function $f(x)$ into two components $f_0(x)$ and $f_1(x)$, and introduces the
modified recurrence relation given as

$$
\begin{aligned}
u_0(x) &= f_0(x), \\
u_1(x) &= f_1(x) + \int_a^b K(x,t)u_0(t)\, dt, \\
u_{n+1}(x) &= \int_a^b K(x,t)u_n(t)\, dt, n \geq 1.
\end{aligned}
$$ (90)

The use of the modified decomposition method minimizes the computations. It is worth noting that a proper selection of $f_0(x)$ and $f_1(x)$ is
essential for a successful use of the modified decomposition method.

The noise terms are defined as the identical terms with opposite signs
that may appear within the components $u_n(x)$, for $n \geq 0$. The noise
terms, if appearing especially within both of the components $u_0(x)$ and
$u_1(x)$, will provide the exact solution by using only the first two iterations.
By canceling the noise terms for $u_0(x)$, the remaining non-canceled terms
of $u_0(x)$ may give the exact solution, and this can be verified through
substitution into the original equation.

In a manner parallel to our earlier analysis, we will investigate three
weakly-singular Fredholm integral equations. We will apply the modified
decomposition method and the noise terms phenomenon when appropriate.

Example 1. Consider the weakly-singular Fredholm integral equation

$$u(x) = 3 + 6\sqrt{x-1} - 6\sqrt{x} + \int_0^1 \frac{u(t)}{\sqrt{x-t}}\, dt, 0 \leq x \leq 1.$$ (91)

We first decompose $f(x)$ into two parts defined as

$$\begin{aligned} f_0(x) &= 3, \\ f_1(x) &= 6\sqrt{x-1} - 6\sqrt{x}. \end{aligned} \quad (92)$$

The modified decomposition method admits the use of the modified recurrence relation as

$$\begin{aligned} u_0(x) &= 3, \\ u_1(x) &= 6\sqrt{x-1} - 6\sqrt{x} + \int_0^1 \frac{u_0(t)}{\sqrt{|x-t|}} \, dt \\ &= 0. \end{aligned} \quad (93)$$

This gives the exact solution by

$$u(x) = 3. \quad (94)$$

Example 2. Consider the weakly-singular Fredholm integral equation

$$u(x) = 1 - x - 2\sqrt{x}(1 - \frac{2}{3}x) - \frac{4}{3}(x-1)^{\frac{3}{2}} + \int_0^1 \frac{u(t)}{\sqrt{x-t}} \, dt, 0 \le x \le 1. \quad (95)$$

We first decompose $f(x)$ into two parts defined as

$$\begin{aligned} f_0(x) &= 1 - x, \\ f_1(x) &= -2\sqrt{x}(1 - \frac{2}{3}x) - \frac{4}{3}(x-1)^{\frac{3}{2}}. \end{aligned} \quad (96)$$

The modified decomposition method admits the use of the modified recurrence relation as

$$\begin{aligned} u_0(x) &= 1 - x, \\ u_1(x) &= -2\sqrt{x}(1 + \frac{2}{3}x) - \frac{4}{3}(x-1)^{\frac{3}{2}} + \int_0^1 \frac{u_0(t)}{\sqrt{|x-t|}} \, dt \\ &= 0. \end{aligned} \quad (97)$$

This gives the exact solution by

$$u(x) = 1 - x. \quad (98)$$

Example 3. Consider the weakly-singular Fredholm integral equation

$$u(x) = 1 + x - \frac{3}{2}x^{\frac{2}{3}} - \frac{9}{10}x^{\frac{5}{3}} + \frac{21}{10}(x-1)^{\frac{2}{3}} + \frac{9}{10}(x-1)^{\frac{5}{3}} + \int_0^1 \frac{u(t)}{(x-t)^{\frac{1}{3}}} \, dt, \quad (99)$$

where $0 \leq x \leq 1$. We first decompose $f(x)$ into two parts defined as

$$
\begin{aligned}
f_0(x) &= 1 + x - \frac{3}{2}x^{\frac{2}{3}}, \\
f_1(x) &= -\frac{9}{10}x^{\frac{5}{3}} + \frac{21}{10}(x-1)^{\frac{2}{3}} + \frac{9}{10}(x-1)^{\frac{2}{3}}.
\end{aligned}
\tag{100}
$$

The modified decomposition method admits the use of the modified recurrence relation as

$$
\begin{aligned}
u_0(x) &= 1 + x - \frac{3}{2}x^{\frac{2}{3}}, \\
u_1(x) &= -\frac{9}{10}x^{\frac{5}{3}} + \frac{21}{10}(x-1)^{\frac{2}{3}} + \frac{9}{10}(x-1)^{\frac{2}{3}} + \int_0^1 \frac{u_0(t)}{\sqrt{|x-t|}}\, dt \\
&= \frac{3}{2}x^{\frac{2}{3}} + \text{other terms.}
\end{aligned}
\tag{101}
$$

Cancelling the noise term from $u_0(x)$ gives the exact solution by

$$
u(x) = 1 + x. \tag{102}
$$

Exercises 6.5

Use the modified decomposition method and the noise terms phenomenon to solve the following weakly-singular Fredholm integral equations of the second kind:

1. $u(x) = x^2 - \frac{16}{15}x^{\frac{5}{2}} + \frac{2\sqrt{x-1}}{5} + \frac{8x\sqrt{x-1}}{15} + \frac{16x^2\sqrt{x-1}}{15} + \int_0^1 \frac{u(t)}{\sqrt{x-t}}u(t)\, dt.$

2. $u(x) = 10 + 20(x-1)^{\frac{1}{2}} - 20(x+1)^{\frac{1}{2}} + \int_{-1}^1 \frac{u(t)}{(x-t)^{\frac{1}{2}}}\, dt.$

3. $u(x) = 10x - 9x^{\frac{5}{3}} + 6(x-1)^{\frac{2}{3}} + 9x(x-1)^{\frac{2}{3}} + \int_0^1 \frac{u(t)}{(x-t)^{\frac{1}{3}}}u(t)\, dt.$

4. $u(x) = 3 + 10x - 9x^{\frac{1}{3}} - \frac{45}{2}x^{\frac{4}{3}} + \frac{33}{2}(x-1)^{\frac{1}{3}} + \frac{45}{2}x(x-1)^{\frac{1}{3}} + \int_0^1 \frac{u(t)}{(x-t)^{\frac{2}{3}}}\, dt.$

5. $u(x) = x + x^2 - \frac{16}{5}x^{\frac{5}{4}}(1 + \frac{8}{9}x) + \frac{8}{45}(x-1)^{\frac{1}{4}}(7 + 22x + 16x^2) + \int_0^1 \frac{u(t)}{(x-t)^{\frac{3}{4}}}\, dt.$

Chapter 7

Nonlinear Fredholm Integral Equations

7.1 Introduction

So far in this text we have been mainly concerned with studying different methods for solving linear integral equations of the first and the second kind. We pointed out earlier that nonlinear integral equations yield a considerable amount of difficulties. However, with the recent methods developed, it seems reasonable to present some reliable and powerful techniques that will make the study of specific cases of nonlinear integral equations successful and valuable. In general, the solution of the nonlinear integral equation is not unique. However, the existence of a solution of nonlinear integral equations with specific conditions is possible. Because we will concern ourselves with nonlinear Fredholm integral equations that will give solutions, therefore we will not discuss in this text the theorem of existence of solutions of nonlinear equations. For more information about the conditions that are necessary for the existence of solutions for nonlinear equations, the reader is advised to look in other texts such as [19] and [44].

The purpose of this chapter is to introduce reliable and easily computable techniques for solving specific cases of nonlinear Fredholm integral equations. As indicated in Chapter 1, given $F(u(t))$ a nonlinear function in $u(t)$, integral equations of the form

$$u(x) = f(x) + \lambda \int_a^b K(x,t)\, F(u(t))dt, \qquad (1)$$

191

and

$$u(x) = f(x) + \lambda \int_0^x K(x,t)\, F(u(t))dt, \tag{2}$$

are called *nonlinear* Fredholm integral equations and *nonlinear* Volterra integral equations respectively. The function $F(u(t))$ is nonlinear in $u(t)$ such as $u^2(t)$, $u^3(t)$, $e^{u(t)}$, and λ is a parameter. However, in this text, we will restrict our discussion to the case where $F(u(t)) = u^n(t)$, $n \geq 2$, whereas other nonlinear integral equations that involve nonlinear terms other than $u^n(t)$ can be handled in a very similar manner. The following are examples of nonlinear Fredholm integral equations:

$$u(x) = 1 + \lambda \int_0^1 u^2(t)dt, \tag{3}$$

$$u(x) = x + \int_0^1 xtu^3(t)dt. \tag{4}$$

7.2   Nonlinear Fredholm Integral Equations of the Second Kind

In this section we will discuss the most successful methods for solving nonlinear Fredholm integral equations of the second kind. Recall that for nonlinear Fredholm integral equations of the second kind, the unknown function $u(x)$ appears inside and outside the integral sign. It has been concluded that the *direct computation method* proved to be reliable in that it handled successfully the linear Fredholm integral equations and the Fredholm integro-differential equations in Chapters 2 and 4 respectively. Based on this conclusion, the direct computation method will be implemented here to provide the exact or (closed form) solution as will be discussed later.

Moreover, the *Adomian decomposition method* proved to be an elegant tool in handling linear and nonlinear equations as well. Accordingly, it is useful to use this method here to obtain the solution in the form of a power series. However, one important fact concerning the decomposition method, in handling the nonlinear problems, is that it requires the use of the so called Adomian polynomials that will represent the nonlinear terms such as $u^n(t)$, $n \geq 2$ or $e^{u(x)}$ that appear under the integral sign. The scheme that to construct the Adomian polynomials will be explained in details later.

7.2.1   The Direct Computation Method

The direct computation method was presented before and used effectively in Chapters 2 and 4. The method approaches the problem directly and gives all possible solutions of the nonlinear equations if the equation has more than one solution.

As stated before we will focus our study on the nonlinear Fredholm integral equations of the second kind of the form

$$u(x) = f(x) + \lambda \int_a^b K(x,t)\, u^n(t) dt, \tag{5}$$

where the kernel $K(x,t)$ will be assumed a separable kernel. Without loss of generality, we may consider the kernel $K(x,t)$ to be expressed by

$$K(x,t) = g(x)h(t). \tag{6}$$

Consequently, we rewrite the equation (5) as

$$u(x) = f(x) + \lambda g(x) \int_a^b h(t)\, u^n(t) dt. \tag{7}$$

We can easily observe that the definite integral in the right hand side of (7) depends only on the variable t. Therefore, we will follow the approach usually used in the *direct computation method*, hence we set

$$\alpha = \int_a^b h(t)\, u^n(t) dt, \tag{8}$$

where the constant α represents the numerical value of the integral. Accordingly, we may rewrite (7) as

$$u(x) = f(x) + \lambda \alpha g(x). \tag{9}$$

Substituting $u(x)$ from (9) into (8), and integrating the easily computable integral yield the numerical value of the constant α. The exact solution $u(x)$ is readily determined upon substituting the obtained value of α in (9).

We point out that the derived solution $u(x)$ in (9) depends on the parameter λ. Accordingly, it is normal to discuss all possible values of λ that will define real solutions for $u(x)$. As a result, two related phenomena, termed as the *bifurcation point* and the *singular point*, may appear. These phenomena have been introduced by [19, 44] and others. For simplicity reasons, the direct computation method and the phenomena of the *bifurcation point* and the *singular point* of the nonlinear integral equation will be illustrated by the following examples.

Example 1. We consider the nonlinear Fredholm integral equation

$$u(x) = 2 + \lambda \int_0^1 u^2(t)\, dt. \tag{10}$$

Setting

$$\alpha = \int_0^1 u^2(t)\, dt, \tag{11}$$

carries (10) into

$$u(x) = 2 + \lambda\alpha. \tag{12}$$

Substituting (12) into (11) yields

$$\alpha = \int_0^1 (2 + \lambda\alpha)^2\, dt, \tag{13}$$

which gives

$$\alpha = (2 + \lambda\alpha)^2, \tag{14}$$

or equivalently

$$\lambda^2\alpha^2 + (4\lambda - 1)\alpha + 4 = 0. \tag{15}$$

Solving the quadratic equation (15) for α gives

$$\alpha = \frac{(1 - 4\lambda) \pm \sqrt{1 - 8\lambda}}{2\lambda^2}, \tag{16}$$

so that substituting (16) into (12) yields

$$u(x) = \frac{1 \pm \sqrt{1 - 8\lambda}}{2\lambda}. \tag{17}$$

Singular Points and Bifurcation Points

It is obvious from (17) that the number of solutions depends on λ. The bifurcation point is a value of λ such that when λ changes though the bifurcation value, then the number of real solutions will change as a result. To explain this, we examine the following possible values of λ:

(i) For $\lambda = 0$, using (10) we obtain $u(x) = 2$, but using (17) we find that $u(x)$ is infinite. For this reason, the point $\lambda = 0$ is called a *singular point* of the equation (10).

(ii) For $\lambda < \frac{1}{8}$, the equation (10) has two real solutions. It is clear in this case that the solution is not unique. This is normal for nonlinear integral equations.

(iii) For $\lambda = \frac{1}{8}$, the equation (10) has one real solution and the point $\lambda = \frac{1}{8}$ is called a *bifurcation point*. The real solution in this case is $u(x) = 4$.

This in turn explains that for $\lambda = \frac{1}{8}$ we obtain one solution, whereas when λ changes to $\lambda < \frac{1}{8}$, such as $\lambda = -1$, the number of real solutions will be changed from one to two, namely $u(x) = -2, 1$.

Example 2. We next consider the nonlinear Fredholm integral equation

$$u(x) = \frac{7}{8}x + \frac{1}{2}\int_0^1 xtu^2(t)\,dt. \tag{18}$$

Setting

$$\alpha = \int_0^1 tu^2(t)\,dt, \tag{19}$$

carries (18) into

$$u(x) = \left(\frac{7}{8} + \frac{1}{2}\alpha\right)x. \tag{20}$$

Substituting (20) into the equation (19) gives

$$\alpha = \int_0^1 t\left(\frac{7}{8} + \frac{1}{2}\alpha\right)^2 t^2\,dt, \tag{21}$$

which gives

$$\alpha = \frac{1}{4}\left(\frac{7}{8} + \frac{1}{2}\alpha\right)^2, \tag{22}$$

or equivalently

$$(4\alpha - 1)(4\alpha - 49) = 0, \tag{23}$$

so that

$$\alpha = \frac{1}{4}, \frac{49}{4}. \tag{24}$$

Accordingly, two real solutions given by

$$u(x) = x, 7x, \tag{25}$$

are obtained upon using (24) into (20).

Example 3. We now consider the nonlinear Fredholm integral equation

$$u(x) = 1 - \frac{51}{35}x^2 + \int_0^1 x^2u^3(t)\,dt. \tag{26}$$

Setting

$$\alpha = \int_0^1 u^3(t)\,dt, \tag{27}$$

carries (26) into

$$u(x) = 1 + (\alpha - \frac{51}{35})x^2. \tag{28}$$

Substituting (28) into the equation (27), and proceeding as before we find

$$\alpha = \frac{16}{35}, \frac{79}{35}, -\frac{89}{35}.$$

(29)

Accordingly, three real solutions given by

$$u(x) = 1 - x^2, 1 - 4x^2, 1 + \frac{4}{5}x^2,$$

(30)

are obtained upon using (29) into (28).

To obtain more than one solution is normal for a nonlinear equation. Linear equations give unique solutions. As will be seen later, the Adomian decomposition method, although is reliable to solve integral equations but will give only one solution, but this will not indicate that nonlinear equation gives only one solution.

Exercises 7.2.1

In exercises 1-5, use the *Direct Computation Method* to solve the given nonlinear integral equations. Also find the *singular point* and the *bifurcation point* of each equation

1. $u(x) = 1 + \frac{1}{2}\lambda \int_0^1 u^2(t)\, dt$

2. $u(x) = 1 - \lambda \int_0^1 u^2(t)\, dt$

3. $u(x) = 1 + \lambda \int_0^1 tu^2(t)\, dt$

4. $u(x) = 1 + \lambda \int_0^1 t^2 u^2(t)\, dt$

5. $u(x) = 1 + \lambda \int_0^1 t^3 u^2(t)\, dt$

In exercises 6-12, use the *Direct Computation Method* to solve the following nonlinear integral equations

6. $u(x) = 2 - \frac{4}{3}x + \int_0^1 xt^2 u^2(t)\, dt$

7. $u(x) = \sin x - \frac{\pi}{8} + \frac{1}{2}\int_0^{\pi/2} u^2(t)\, dt$

8. $u(x) = \cos x - \frac{\pi}{8} + \frac{1}{2}\int_0^{\pi/2} u^2(t)\, dt$

9. $u(x) = x - \frac{1}{8} + \frac{1}{2}\int_0^1 tu^2(t)\, dt$

10. $u(x) = x^2 - \dfrac{1}{10} + \dfrac{1}{2} \displaystyle\int_0^1 u^2(t)\, dt$

11. $u(x) = x - \dfrac{5}{6} + \displaystyle\int_0^1 \left(u(t) + u^2(t)\right)\, dt$

12. $u(x) = x - 1 + \dfrac{3}{4} \displaystyle\int_0^1 \left(2t + u^2(t)\right)\, dt$

7.2.2   The Adomian Decomposition Method

The Adomian decomposition method introduces a reliable analysis to handle the nonlinear integral equations. The method provides a rapidly convergent series solution without using any restrictive assumptions such as linearization or perturbation assumptions. The linear term $u(x)$ in the equation is usually expressed in a power series in a similar manner to that discussed before for linear integral equations. However, an essential scheme is required for representing the nonlinear term $u^n(x)$ involved in the equation. Moreover, the decomposition method does not investigate the existence of the solution of the problem. In addition, it gives one solution although nonlinear equations generally give more than one solution.

In the following, the decomposition method will be fully discussed for nonlinear equations [1] and [2]. For simplicity reasons, we consider the simple form of nonlinear Fredholm integral equation

$$u(x) = f(x) + \lambda \int_a^b K(x,t)\, u^n(t)dt, \qquad (31)$$

where other forms of nonlinearity of $F(u(t))$ can be handled in a parallel manner. The solution $u(t)$ of (31) can be represented normally by the decomposition series

$$u(x) = \sum_{n=0}^{\infty} u_n(x), \qquad (32)$$

where the components $u_n(x)$, $n \geq 0$ can be computed in a recursive manner as discussed before. However, as stated above the nonlinear term $u^n(t)$ of the equation (31) should be represented, using a distinct scheme, by the so called Adomian polynomials $A_n(t)$.

Adomian Polynomials

The simple and practical scheme that will construct Adomian polynomials begins by assuming that the nonlinear term $u^n(t)$ under the integral sign in (31) will be equated to the polynomial series

$$u^n(t) = \sum_{n=0}^{\infty} A_n(t), \qquad (33)$$

where the $A_n(t)$ are the so called Adomian polynomials. It was formally proved by [1–5] that the Adomian polynomials can be completely determined by using the following scheme

$$
\begin{aligned}
A_0 &= F(u_0), \\
A_1 &= u_1 \, F'(u_0), \\
A_2 &= u_2 \, F'(u_0) + \frac{u_1^2}{2!} \, F''(u_0), \\
A_3 &= u_3 \, F'(u_0) + u_1 u_2 \, F''(u_0) + \frac{u_1^3}{3!} \, F'''(u_0), \\
A_4 &= u_4 \, F'(u_0) + \left(\frac{1}{2!}u_2^2 + u_1 u_3\right) F''(u_0) \\
&\quad + \frac{u_1^2 u_2}{2!} \, F'''(u_0) + \frac{1}{4!}u_1^4 F^{(iv)}(u_0),
\end{aligned}
\qquad (34)
$$

$$\cdots$$

where $F(u(t))$ is the nonlinear function, and in this specific equation it is given by $F(u(t)) = u^n(t)$. In addition we point out that A_0 depends only on u_0, A_1 depends only on u_0 and u_1, A_2 depends only on u_0, u_1 and u_2, etc. It is to be noted that the sum of the subscripts of each term of A_n is equal to n. For example, in A_2, we the two terms include u_2 and $u_1^2 = u_1 u_1$, where the sum of subscripts of each term is 2 and 2. For A_3, the terms are u_3, $u_1 u_2$, and u_1^3, with sum of subscripts is 3 for each term. More details about the derivation of Adomian polynomials can be found in [1–4, 44].

It is remarked before that in this section, for simplicity reasons, we will discuss nonlinear Fredholm integral equation (31) where the nonlinear term is of the form $F(u(t)) = u^n(t)$ only.

In the following we explain how we can use the scheme given by (34) to define Adomian polynomials:
(i) Consider the nonlinear function

$$F(u) = u^2(x), \qquad (35)$$

then

$$
\begin{cases}
A_0 &= u_0^2, \\
A_1 &= 2u_0 u_1 \\
A_2 &= 2u_0 u_2 + u_1^2 \\
A_3 &= 2u_0 u_3 + 2u_1 u_2 \\
\vdots &
\end{cases}
\qquad (36)
$$

We can easily observe from (36) that in the polynomial A_2 the sum of the subscripts of each term of the two terms $2u_0u_2$ and u_1u_1 is equal to the subscript of A_2. The same fact holds for other polynomials.

(ii) For the nonlinear function

$$F(u) = u^3 \tag{37}$$

we find

$$\begin{cases} A_0 = u_0^3 \\ A_1 = 3u_0^2u_1 \\ A_2 = 3u_0^2u_2 + 3u_0u_1^2 \\ A_3 = 3u_0^2u_3 + 6u_0u_1u_2 + u_1^3 \\ \vdots \end{cases} \tag{38}$$

(iii) For the nonlinear function

$$F(u) = u^4 \tag{39}$$

we find

$$\begin{cases} A_0 = u_0^4 \\ A_1 = 4u_0^3u_1 \\ A_2 = 4u_0^3u_2 + 6u_0^2u_1^2 \\ A_3 = 4u_0^3u_3 + 12u_0^2u_1u_2 + 4u_0u_1^3 \\ \vdots \end{cases} \tag{40}$$

(iv) For the nonlinear function

$$F(u) = e^u \tag{41}$$

we use the scheme (34) to generate the Adomian polynomials

$$\begin{cases} A_0 = e^{u_0} \\ A_1 = u_1e^{u_0} \\ A_2 = \left(\dfrac{u_1^2}{2} + u_2 \right) e^{u_0} \\ A_3 = \left(\dfrac{u_1^3}{6} + u_1u_2 + u_3 \right) e^{u_0} \\ \vdots \end{cases} \tag{42}$$

(v) For the nonlinear function

$$F(u) = \sin u(x) \tag{43}$$

we use the scheme (34) to generate the Adomian polynomials

$$\left\{ \begin{array}{rl} A_0 &= \sin u_0 \\ A_1 &= u_1 \cos u_0 \\ A_2 &= u_2 \cos u_0 - \dfrac{1}{2!} u_1^2 \sin u_0 \\ \vdots & \end{array} \right. \tag{44}$$

(vi) For the nonlinear function

$$F(u) = \cos u(x) \tag{45}$$

we use the scheme (34) to generate the Adomian polynomials

$$\left\{ \begin{array}{rl} A_0 &= \cos u_0 \\ A_1 &= u_1 \sin u_0 \\ A_2 &= -u_2 \sin u_0 - \dfrac{1}{2!} u_1^2 \cos u_0 \\ \vdots & \end{array} \right. \tag{46}$$

The last three examples, where the nonlinear functions are $e^u, \sin u, \cos u$, have been introduced for further studies beyond the scope of this text. These examples are presented here for illustration purposes only. As stated before, $F(u)$ may have other nonlinear forms such as $\sinh(u), \cosh(u), \ln u$. However, our concern in this text will be on nonlinear functions of the form $F(u) = u^n$. We point out that for, $n \geq 3$, the decomposition method is easier to use than the direct computation method. In the latter case, an algebraic equation of higher degree is obtained.

We now return to the main goal of our discussion to determine the components $u_0(x), u_1(x), u_2(x), \ldots$ of the solution $u(x)$. This can be done by substituting the decomposition (32) that represents the linear term $u(x)$, and the decomposition (33) that represents the nonlinear term $u^n(x)$ into (31). Hence we obtain

$$\sum_{n=0}^{\infty} u_n(x) = f(x) + \lambda \int_a^b K(x,t) \left(\sum_{n=0}^{\infty} A_n(t) \right) dt, \tag{47}$$

or simply

$$u_0(x) + u_1(x) + u_2(x) + \cdots$$
$$= f(x) + \lambda \int_a^b K(x,t) \left[A_0(t) + A_1(t) + A_2(t) + \cdots \right] dt. \tag{48}$$

The components $u_0(x), u_1(x), u_2(x), \ldots$ are completely determined by using the recurrence relation

$$
\begin{cases}
u_0(x) = f(x), \\[2mm]
u_1(x) = \lambda \displaystyle\int_a^b K(x,t)A_0(t)dt, \\[2mm]
u_2(x) = \lambda \displaystyle\int_a^b K(x,t)A_1(t)dt, \\[2mm]
\vdots \\[2mm]
u_{n+1}(x) = \lambda \displaystyle\int_a^b K(x,t)A_n(t)dt, \quad n \geq 0.
\end{cases}
\tag{49}
$$

Consequently the solution of (31) in a series form is immediately determined. As indicated earlier, the series obtained may converge to the exact solution in a closed form, or a truncated series $\sum_{n=1}^{k} u_n(x)$ may be used if a numerical approximation is desired. It is worth noting that the convergence question of the method was addressed by many authors.

Before we give a clear view of the method that handles the occurrence of the nonlinear term $u^n(x)$, it is useful to discuss the following remarks:

(i) Even though the decomposition method gives only one solution for each nonlinear equation, this does not indicate the uniqueness of the solution of the nonlinear integral equations. We have seen that two real solutions were obtained for examples discussed above by using the direct computation method, and a unique solution is determined under specific conditions only. This is consistent with the fact that the decomposition method does not address the existence and the uniqueness concept.

(ii) The modified decomposition method, that was introduced before and the criteria of the self-cancelling noise terms can be implemented here to speed the process of obtaining the solution.

(iii) It is important to emphasize that Adomian polynomials A_n can be calculated for complicated nonlinearities of $F(u)$.

(iv) There are other techniques that can be used to construct Adomian polynomials, but we will use the Adomian algorithm that we introduced earlier in this text.

In the following we will outline a brief framework of the *modified decomposition method*. Splitting the nonhomogeneous part $f(x)$ into two parts

$f_0(x)$ and $f_1(x)$ enables us to follow the scheme given by

$$
\begin{cases}
u_0(x) = f_0(x), \\[2mm]
u_1(t) = f_1(x) + \lambda \displaystyle\int_a^b K(x,t)A_0(t)dt, \\[2mm]
u_2(t) = \lambda \displaystyle\int_a^b K(x,t)A_1(t)dt, \\[2mm]
\vdots \\[2mm]
u_{n+1}(x) = \lambda \displaystyle\int_a^b K(x,t)A_n(t)dt, \quad n \geq 0.
\end{cases}
\tag{50}
$$

Having determined the components $u_0(x), u_1(x), u_2(x), \ldots$ leads to the solution in a series form upon using (32).

The decomposition method and the modified decomposition method will be illustrated by studying the following examples. For comparison reasons, we discuss Examples 1 and 2 above that were solved before by using the direct computation method.

Example 1. We consider the nonlinear Fredholm integral equation

$$
u(x) = 2 + \lambda \int_0^1 u^2(t)\, dt, \quad \lambda \leq \frac{1}{8}.
\tag{51}
$$

In this example we have

$$
F(u) = u^2(x),
\tag{52}
$$

which generates the following polynomials

$$
\begin{cases}
A_0 = u_0^2, \\[2mm]
A_1 = 2u_0u_1 \\[2mm]
A_2 = 2u_0u_2 + u_1^2 \\[2mm]
A_3 = 2u_0u_3 + 2u_1u_2 \\[2mm]
\vdots
\end{cases}
\tag{53}
$$

Using the recursive algorithm (49) we find

$$u_0(x) = 2, \tag{54}$$

$$u_1(x) = \lambda \int_0^1 A_0(t)dt$$
$$= 4\lambda, \tag{55}$$

$$u_2(x) = \lambda \int_0^1 A_1(t)dt$$
$$= 16\lambda^2, \tag{56}$$

$$u_3(x) = \lambda \int_0^1 A_2(t)dt$$
$$= 80\lambda^3, \tag{57}$$

and so on. The solution in a series form is then given by

$$u(x) = 2 + 4\lambda + 16\lambda^2 + 80\lambda^3 + \cdots. \tag{58}$$

It is obvious from (58) that only one solution has been obtained by using the decomposition method. We recall that two answers have been obtained earlier by using the direct computation method given by

$$u(x) = \frac{1 \pm \sqrt{1 - 8\lambda}}{2\lambda}, \tag{59}$$

which, by using the binomial theorem to expand the square root, gives

$$u(x) = \frac{1 \pm \left(1 - 4\lambda - 8\lambda^2 - 32\lambda^3 - 160\lambda^4 + \cdots\right)}{2\lambda}, \tag{60}$$

so that $u(x)$ has the two expansions

$$u(x) = \frac{1}{\lambda} - \left(2 + 4\lambda + 16\lambda^2 + 80\lambda^3 + \cdots\right), \tag{61}$$

and

$$u(x) = \left(2 + 4\lambda + 16\lambda^2 + 80\lambda^3 + \cdots\right). \tag{62}$$

We can easily observe that the solution (58) obtained by using the decomposition method is consistent with the second solution (62) obtained by using the direct computation method. However, the decomposition method did not address any procedure to find the second solution (61).

Example 2.   We next consider the nonlinear Fredholm integral equation

$$u(x) = \frac{7}{8}x + \frac{1}{2}\int_0^1 xtu^2(t)\,dt. \tag{63}$$

In this example the Adomian polynomials for the nonlinear term

$$F(u) = u^2(x), \tag{64}$$

are given by

$$\left\{\begin{array}{rcl} A_0 & = & u_0^2, \\ A_1 & = & 2u_0u_1 \\ A_2 & = & 2u_0u_2 + u_1^2 \\ A_3 & = & 2u_0u_3 + 2u_1u_2 \\ & \vdots & \end{array}\right. \tag{65}$$

Using the recursive algorithm (49) we find

$$u_0(x) = \frac{7}{8}x, \tag{66}$$

$$u_1(x) = \frac{1}{2}x\int_0^1 tA_0(t)dt,$$

$$= \frac{49}{512}x, \tag{67}$$

$$u_2(x) = \frac{1}{2}x\int_0^1 tA_1(t)dt,$$

$$= \frac{343}{16384}x, \tag{68}$$

and so on. The solution in a series form is given by

$$\begin{array}{rcl} u(x) & = & \dfrac{7}{8}x + \dfrac{49}{512}x + \dfrac{343}{16384}x + \cdots, \\ & = & 0.875x + 0.0957031x + 0.02935x + \cdots, \\ & \simeq & x. \end{array} \tag{69}$$

Note that we can use the modified decomposition method, where we set $f_0(x) = x$ and $f_1(x) = -\frac{1}{8}x$ and proceed as before. We remark that two solutions $u(x) = x$ and $u(x) = 7x$ were obtained in (25) by using the direct computation method.

Example 3.   As a third example we consider the nonlinear Fredholm integral equation

$$u(x) = 1 - \frac{1}{3}x + \int_0^1 xt^2u^3(t)\,dt. \tag{70}$$

The Adomian polynomials for the nonlinear term

$$F(u) = u^3(x), \tag{71}$$

have been determined before in (38). In this example we will use the *modified decomposition method*. Setting

$$u_0(x) = 1 \tag{72}$$

leads to

$$\begin{aligned} u_1(x) &= -\frac{1}{3}x + x \int_0^1 t^2 A_0(t) dt = 0, \\ u_k(x) &= 0, \quad k \geq 1. \end{aligned} \tag{73}$$

This yields the exact solution

$$u(x) = 1. \tag{74}$$

For comparison reasons, we will solve this integral equation by using the direct computation method. From (70) we set

$$u(x) = 1 + (\alpha - \frac{1}{3})x, \tag{75}$$

where

$$\alpha = \int_0^1 t^2 u^3(t) \, dt. \tag{76}$$

To determine α, we substitute (75) into (76) where we get an algebraic equation of third degree in α given by

$$\alpha = \frac{1}{6}\alpha^3 + \frac{13}{30}\alpha^2 + \frac{73}{180}\alpha + \frac{233}{1620}. \tag{77}$$

Solving this equation, by using a calculator or a computer software as Maple, we find

$$\alpha = \frac{1}{3}, -\frac{22}{15} + \frac{\sqrt{474}}{10}, -\frac{22}{15} - \frac{\sqrt{474}}{10}. \tag{78}$$

Substituting these values of α in (75) gives the three distinct solutions given by

$$u(x) = 1, \ 1 - (\frac{9}{5} - \frac{\sqrt{474}}{10})x, \ 1 - (\frac{9}{5} + \frac{\sqrt{474}}{10})x. \tag{79}$$

This example shows that the direct computation method produces all three solutions, whereas the decomposition method gives only one of these three solutions. Although the Adomian decomposition method is effective and

reliable, but because it gives only one solution for a nonlinear integral equation, it is considered by many as an aspect of weakness in solving nonlinear
integral equations.

Exercises 7.2.2

Use the *decomposition method* or the *modified decomposition method* to solve
the following nonlinear Fredholm integral equations:

1. $u(x) = 1 + \lambda \int_0^1 t u^2(t)\, dt, \quad \lambda \le \dfrac{1}{2}.$

2. $u(x) = 1 + \lambda \int_0^1 t^3 u^2(t)\, dt, \quad \lambda \le 1.$

3. $u(x) = 2\sin x - \dfrac{\pi}{8} + \dfrac{1}{8} \int_0^{\pi/2} u^2(t)\, dt\ .$

4. $u(x) = 2\cos x - \dfrac{\pi}{8} + \dfrac{1}{8} \int_0^{\pi/2} u^2(t)\, dt.$

5. $u(x) = \sec x - x + x \int_0^{\pi/4} u^2(t)\, dt.$

6. $u(x) = \dfrac{3}{2}x + \dfrac{3}{8} \int_0^1 x u^2(t)\, dt.$

7. $u(x) = x^2 - \dfrac{1}{12} + \dfrac{1}{2} \int_0^1 t u^2(t)\, dt.$

8. $u(x) = x - \dfrac{\pi}{8} + \dfrac{1}{2} \int_0^1 \dfrac{1}{1 + u^2(t)}\, dt.$

9. $u(x) = x - 1 + \dfrac{2}{\pi} \int_{-1}^1 \dfrac{1}{1 + u^2(t)}\, dt.$

10. $u(x) = x - \dfrac{1}{4}\ln 2 + \dfrac{1}{2} \int_0^1 \dfrac{t}{1 + u^2(t)}\, dt.$

11. $u(x) = \sin x + \cos x - \dfrac{\pi + 2}{8} + \dfrac{1}{4} \int_0^{\pi/2} u^2(t)\, dt.$

12. $u(x) = \sinh x - 1 + \int_0^1 \left(\cosh^2(t) - u^2(t)\right)\, dt.$

13. $u(x) = \cos x + 2 - \int_0^1 \left(1 + \sin^2(t) + u^2(t)\right)\, dt.$

14. $u(x) = \sec x - x + \int_0^1 x \left(u^2(t) - \tan^2(t)\right)\, dt.$

7.2.3 The Variational Iteration Method

In this section we will present the *variational iteration method* for solving nonlinear Fredholm integral equations. We will follow a manner parallel to the analysis presented earlier in this text. This means that to use this method for solving integral equations, linear or nonlinear, we first should convert the integral equation to its equivalent differential equation, or to its equivalent integro-differential equation.

In this chapter, we will consider the kernel $K(x, t)$ to be separable of the form $K(x, t) = g(x)h(t)$. The Fredholm integral equation can be converted to an identical Fredholm integro-differential equation by differentiating both sides, where an initial condition should also be derived. For simplicity, we will study only the cases where $h(x) = x^n, n \geq 1$. In what follows we will present the main steps for using this method.

The standard Fredholm integral equation is of the form

$$u(x) = f(x) + \int_a^b K(x, t) F(u(t)) dt, \tag{80}$$

or equivalently

$$u(x) = f(x) + g(x) \int_a^b h(t) F(u(t)) dt, \quad K(x, t) = g(x)h(t), \tag{81}$$

where $F(u(t))$ is a nonlinear function of $u(t)$. Recall that the integral at the right side of this equation depends on t only, hence it is equivalent to a constant. Differentiating both sides of (81) with respect to x gives

$$u'(x) = f'(x) + g'(x) \int_a^b h(t) F(u(t)) dt. \tag{82}$$

The variational iteration method admits the use of a correction functional for the integro-differential equation in the form

$$u_{n+1}(x) = u_n(x) + \int_0^x \lambda(\xi) \left(u_n'(\xi) - f'(\xi) - g'(\xi) \int_a^b h(r) F(u_n(r)) \, dr \right) d\xi \tag{83}$$

where λ is a general Lagrange multiplier. A list of some of these Lagrange multipliers was formally derived and given in Section 2.3. Having determined λ, an iteration formula, can be constructed using the correction functional that will allow us to determine the successive approximations $u_{n+1}(x), n \geq 0$ of the solution $u(x)$. Notice that $u_n(x)$ gives the successive approximations of the solution and not the components as in the case when Adomian method is used. The significant feature of the variational iteration method comes from its power to approach the nonlinear problems directly in a straightforward manner without any need to use Adomian polynomials as in the Adomian decomposition method. This in turn leads to minimizing the computational work.

The zeroth approximation u_0 can be any selective function. However, using the given initial value $u(0)$ is preferably used for the selective zeroth approximation u_0 as will be seen later. Consequently, the solution is given by

$$u(x) = \lim_{n \to \infty} u_n(x). \tag{84}$$

The variational iteration method will be illustrated by studying the following nonlinear Fredholm integral equations.

Example 1. Use the variational iteration method to solve the nonlinear Fredholm integral equation

$$u(x) = \frac{7}{8}x + \frac{1}{2} \int_0^1 xtu^2(t)dt. \tag{85}$$

Differentiating both sides of this equation with respect to x yields

$$u'(x) = \frac{7}{8} + \frac{1}{2} \int_0^1 tu^2(t)dt, \, u(0) = 0. \tag{86}$$

The correction functional for this equation is given by

$$u_{n+1}(x) = u_n(x) - \int_0^x \left(u_n'(\xi) - \frac{7}{8} - \frac{1}{2} \int_0^1 ru_n^2(r)\, dr \right)\, d\xi, \tag{87}$$

where we used $\lambda = -1$ for first-order integro-differential equations. It is preferable to select $u_0(x) = u(0) = 0$. Using this selection into the correction functional gives the following successive approximations

$$
\begin{aligned}
u_0(x) &= 0, \\
u_1(x) &= \frac{7}{8}x = 0.875x, \\
u_2(x) &= \frac{497}{512}x = 0.9707x, \\
u_3(x) &= \frac{2082017}{2097152}x = 0.9928x, \\
u_4(x) &= \frac{35121120366017}{35184372088832}x = 0.9982x, \\
&\vdots
\end{aligned}
\tag{88}
$$

The VIM admits the use of

$$u(x) = \lim_{n \to \infty} u_n(x) = x. \tag{89}$$

It is worth pointing out that the variational iteration method gives only one solution for this nonlinear equation, although the equation is a nonlinear equation. However, this equation has two solutions as obtained by the direct computation method, where we found

$$u(x) = x, 7x. \tag{90}$$

Example 2. Use the variational iteration method to solve the nonlinear Fredholm integral equation

$$u(x) = 1 - \frac{51}{35}x^2 + \int_0^1 x^2 u^3(t)dt. \tag{91}$$

Differentiating both sides of this equation with respect to x yields

$$u'(x) = \frac{102}{35}x + 2x \int_0^1 u^3(t)dt, u(0) = 1. \tag{92}$$

The correction functional for this equation is given by

$$u_{n+1}(x) = u_n(x) - \int_0^x \left(u_n'(\xi) - \frac{102}{35}\xi - 2\xi \int_0^1 u_n^3(r)\, dr \right) d\xi, \tag{93}$$

where we used $\lambda = -1$ for first-order integro-differential equations. Proceeding as before, we obtain the following successive approximations

$$
\begin{aligned}
u_0(x) &= 1, \\
u_1(x) &= 1 - 0.4571x^2, \\
u_2(x) &= 1 - 0.8025x^2, \\
u_3(x) &= 1 - 0.9471x^2, \\
u_4(x) &= 1 - 0.9874x^2, \\
&\vdots
\end{aligned} \tag{94}
$$

The VIM admits the use of

$$u(x) = \lim_{n \to \infty} u_n(x) = 1 - x^2. \tag{95}$$

However, this equation has three solutions as obtained by the direct computation method, where we found

$$u(x) = 1 - x^2, 1 - 4x^2, 1 + \frac{4}{5}x^2. \tag{96}$$

Example 3. Use the variational iteration method to solve the nonlinear Fredholm integral equation

$$u(x) = 1 + 2x - \frac{33}{10}x^2 + \int_0^1 x^2 t u^2(t) dt. \tag{97}$$

Differentiating both sides of this equation with respect to x yields

$$u'(x) = 2 - \frac{33}{5}x + 2x \int_0^1 t u^2(t) dt, u(0) = 1. \tag{98}$$

Proceeding as before, we obtain the following successive approximations

$$
\begin{aligned}
u_0(x) &= 1, \\
u_1(x) &= 1 + 2x - \frac{14}{5}x^2, \\
u_2(x) &= 1 + 2x - \frac{14}{5}x^2, \\
u_3(x) &= 1 + 2x - \frac{14}{5}x^2, \\
&\vdots
\end{aligned}
\tag{99}
$$

The VIM admits the use of

$$u(x) = \lim_{n \to \infty} u_n(x) = 1 + 2x - \frac{14^2}{5}. \tag{100}$$

However, this nonlinear equation has two solutions given by

$$u(x) = 1 + 2x - \frac{14}{5}x^2, 1 + 2x + x^2. \tag{101}$$

Exercises 7.2.3

Use the *variational iteration method* to solve the following nonlinear Fredholm integral equations:

1. $u(x) = \frac{3}{4}x + \int_0^1 x t u^2(t) \, dt.$

2. $u(x) = x^2 - \frac{1}{6}x + \int_0^1 x t u^2(t) \, dt.$

3. $u(x) = x^2 - \frac{1}{8}x + \int_0^1 x t u^2(t) \, dt.$

4. $u(x) = x - \frac{1}{5}x^2 + \int_0^1 x^2 t^2 u^2(t) \, dt.$

5. $u(x) = x + \frac{34}{105}x^2 + \int_0^1 x^2 t^2 u^2(t) \, dt.$

7.3 Nonlinear Fredholm Integral Equations of the First Kind

The nonlinear Fredholm integral equations of the first kind reads

$$f(x) = \int_a^b K(x,t)F(u(t))\,dt, \tag{102}$$

where the kernel $K(x,t)$, which is a function of x and t, and the function $f(x)$ are given real-valued functions. The function $F(u(x))$ is a nonlinear function of $u(x)$, such as $u^2(x), u^3(x), e^{u(x)}$, etc. The linear Fredholm integral equation of the first kind is presented in Chapter 2.

The nonlinear Fredholm integral equation of the first kind was examined in the literature by using many methods, analytical and numerical as well. Some of the methods that are used are the homotopy perturbation method, the Taylor series method, the Galerkin method, the collocation method and others. The methods that we used so far in this text cannot handle this kind of equations independently if it is expressed in its standard form (102). We aim to make our approach to be consistent with our approach for linear Fredholm integral equations. In view of this, we will concern ourselves on using the method of regularization that we applied before in Chapter 2. This means that we should transform the nonlinear form of the integral equations as we applied before.

To determine a solution for the nonlinear Fredholm integral equation of the first kind (102), we first transform it to a linear Fredholm integral equation of the first kind as

$$f(x) = \int_a^b K(x,t)v(t)\,dt, x \in D \tag{103}$$

by using the transformation

$$v(x) = F(u(x)). \tag{104}$$

Assuming that $F(u(x))$ is invertible leads to

$$u(x) = F^{-1}(v(x)). \tag{105}$$

Recall that in Chapter 2, we presented an important remark that the function $f(x)$ must lie in the range of the kernel $K(x,t)$. This means that if we set the kernel by

$$K(x,t) = \cos x \, e^t, \tag{106}$$

then for any integrable function $F(u(x))$ in (102), the resulting $f(x)$ must clearly be a multiple of $\cos x$. However, if $f(x)$ is not a multiple of the x

component, $\cos x$, of the kernel, then we cannot find a solution for (102). This necessary condition on $f(x)$ can be generalized. This means that the data function $f(x)$ must involve components which are matched by the corresponding x components of the kernel $K(x, t)$.

Hadamard [13] introduced a definition to the mathematical term well-posed problem. He believed that well-posed models of physical phenomena should have the following three properties:

1. A solution exists.
2. A solution is unique.
3. Continuous dependence of the solution $u(x)$ on the data $f(x)$.

Examples of well-posed problems include the Dirichlet problem for Laplace's equation, and the heat equation with specified initial conditions. However, models that are not well-posed in the sense of Hadamard are termed ill-posed problems. The nonlinear Fredholm integral equation of the first kind is considered ill-posed problem because it does not satisfy the aforementioned three properties. For any ill-posed problem, a very small change on the data $f(x)$ can lead to a change in the solution $u(x)$.

As stated earlier, we will apply the method of *regularization* that received a considerable amount of interest to confirm its reliability. In what follows we will present a brief summary of the method of regularization that will be used to handle the nonlinear Fredholm integral equations of the first kind.

7.3.1　The Method of Regularization

The method of regularization was established independently by Phillips [33] and Tikhonov [40]. The method was used before in handling linear Fredholm integral equations of the first kind. The method of regularization consists of replacing ill-posed problem by well-posed problem. The method of regularization transforms the linear Fredholm integral equation of the first kind

$$f(x) = \int_a^b K(x, t)v(t)\, dt, x \in D, \tag{107}$$

to an approximation Fredholm integral equation

$$\epsilon v_\epsilon(x) = f(x) - \int_a^b K(x, t)v_\epsilon(t)\, dt, x \in D, \tag{108}$$

where ϵ is a small positive parameter. The resulting equation (108) is a Fredholm integral equation of the second kind rewritten in the form

$$v_\epsilon(x) = \frac{1}{\epsilon}f(x) - \frac{1}{\epsilon}\int_a^b K(x, t)v_\epsilon(t)\, dt, x \in D. \tag{109}$$

Moreover, it was proved in [33, 40] that the solution v_ϵ of equation (109) converges to the solution $v(x)$ of (107) as $\epsilon \to 0$. Having converted the Fredholm integral equation of the first kind to an equivalent Fredholm integral equation of the second kind, we then can use any of the methods that we used earlier in Chapter 2, such as the Adomian decomposition method, the direct computation method, or others. Note that we should convert the nonlinear equation to a linear equation by using a proper transformation as will be seen later. The exact solution $v(x)$ of (107) can thus be obtained by

$$v(x) = \lim_{\epsilon \to 0} v_\epsilon(x). \tag{110}$$

In what follows we will present three illustrative examples where we will use the method of regularization to transform the first kind integral equation to a second kind integral equation. The resulting equation will be solved by any appropriate method that we used before.

Example 1. Use the regularization method and the direct computation method to solve the nonlinear Fredholm integral equation of the first kind

$$e^x = \int_0^1 e^{x-3t} u^3(t) \, dt. \tag{111}$$

We use the transformation

$$v(x) = u^3(x), u(x) = \sqrt[3]{v(x)}, \tag{112}$$

to convert (111) into

$$e^x = \int_0^1 e^{x-3t} v(t) \, dt. \tag{113}$$

Using the method of regularization, Eq. (113) can be transformed to

$$v_\epsilon(x) = \frac{1}{\epsilon} e^x - \frac{1}{\epsilon} \int_0^1 e^{x-3t} v_\epsilon(t) \, dt, \tag{114}$$

that can be rewritten as

$$v_\epsilon(x) = (\frac{1}{\epsilon} - \frac{\alpha}{\epsilon}) e^x, \tag{115}$$

where

$$\alpha = \int_0^1 e^{-3t} v_\epsilon(t) \, dt. \tag{116}$$

To determine α, we substitute (115) into (116) to find

$$\alpha = (\frac{1}{\epsilon} - \frac{\alpha}{\epsilon}) \int_0^1 e^{-2t} \, dt, \tag{117}$$

so that

$$\alpha = \frac{1 - e^{-2}}{1 - e^{-2} + 2\epsilon}. \tag{118}$$

This in turn gives

$$v_\epsilon(x) = \frac{1}{\epsilon}\left(1 - \frac{1 - e^{-2}}{1 - e^{-2} + 2\epsilon}\right)e^x. \tag{119}$$

The exact solution $v(x)$ of (114) can be obtained by

$$v(x) = \lim_{\epsilon \to 0} v_\epsilon(x) = \frac{2e^x}{1 - e^{-2}}. \tag{120}$$

Using (112) gives the exact solution of (111) by

$$u(x) = \sqrt[3]{\frac{2e^x}{1 - e^{-2}}}. \tag{121}$$

One more solution to Eq. (111) is given by

$$u(x) = e^x. \tag{122}$$

Example 2. Use the regularization method and the direct computation method to solve the nonlinear Fredholm integral equation of the first kind

$$\frac{43}{10}x = \int_0^1 xt\, u^4(t)\, dt. \tag{123}$$

We use the transformation

$$v(x) = u^4(x), u(x) = \pm\sqrt[4]{v(x)}, \tag{124}$$

to convert (123) into

$$\frac{43}{10}x = \int_0^1 xt\, v(t)\, dt. \tag{125}$$

Using the method of regularization, Eq. (125) can be transformed to

$$v_\epsilon(x) = \frac{1}{\epsilon}(\frac{43}{10}x) - \frac{1}{\epsilon}\int_0^1 xt v_\epsilon(t)\, dt, \tag{126}$$

that can be rewritten as

$$v_\epsilon(x) = \frac{1}{\epsilon}(\frac{43}{10} - \alpha)x, \tag{127}$$

where

$$\alpha = \int_0^1 t v_\epsilon(t)\, dt. \tag{128}$$

Proceeding as before, we find

$$\alpha = \frac{43}{10 + 10\epsilon}. \tag{129}$$

This in turn gives

$$v_\epsilon(x) = \frac{1}{\epsilon}(\frac{43}{10} - \frac{43}{10 + 10\epsilon})x. \tag{130}$$

The exact solution $v(x)$ of (126) can be obtained by

$$v(x) = \lim_{\epsilon \to 0} v_\epsilon(x) = \frac{129}{10}x. \tag{131}$$

Using (124) gives the exact solution of (123) by

$$u(x) = \pm\sqrt[4]{\frac{129}{10}}x. \tag{132}$$

One more solution to Eq. (123) is given by

$$u(x) = \pm(1 + x). \tag{133}$$

Example 3. Use the regularization method and the direct computation method to solve the nonlinear Fredholm integral equation of the first kind

$$\frac{2}{27}x^2 = \int_0^1 x^2 t^2\, u^2(t)\, dt. \tag{134}$$

We use the transformation

$$v(x) = u^2(x), u(x) = \pm\sqrt{v(x)}, \tag{135}$$

to convert (134) into

$$\frac{2}{27}x^2 = \int_0^1 x^2 t^2\, v(t)\, dt. \tag{136}$$

Using the method of regularization, Eq. (136) can be transformed to

$$v_\epsilon(x) = \frac{1}{\epsilon}(\frac{2}{27}x^2) - \frac{1}{\epsilon}\int_0^1 x^2 t^2 v_\epsilon(t)\, dt, \tag{137}$$

that can be rewritten as

$$v_\epsilon(x) = \frac{1}{\epsilon}(\frac{2}{27} - \alpha)x^2, \tag{138}$$

where

$$\alpha = \int_0^1 t^2 v_\epsilon(t)\, dt. \tag{139}$$

Proceeding as before, we find

$$\alpha = \frac{2}{27(1+5\epsilon)}. \tag{140}$$

This in turn gives

$$v_\epsilon(x) = \frac{1}{\epsilon}\Big(\frac{2}{27} - \frac{2}{27(1+5\epsilon)}\Big)x^2. \tag{141}$$

Following the discussion presented earlier gives the exact solution of (134) by

$$u(x) = \pm\sqrt{\frac{10}{27}}\, x. \tag{142}$$

Two more solutions to Eq. (134) is given by

$$u(x) = \pm \ln x. \tag{143}$$

Exercises 7.3

Combine the regularization method with any method to solve the nonlinear Fredholm integral equations of the first kind

1. $\dfrac{49}{60}x = \displaystyle\int_0^1 xt\, u^2(t)\, dt$

2. $\dfrac{1}{120}x^2 = \displaystyle\int_0^1 x^2 t^2\, u^3(t)\, dt$

3. $e^{2x} = \displaystyle\int_0^1 e^{2x-6t}\, u^3(t)\, dt$

4. $e^x = \displaystyle\int_0^1 e^{x-4t}\, u^4(t)\, dt$

5. $\dfrac{1}{4}x^2 = \displaystyle\int_0^1 x^2 t\, u^2(t)\, dt$

6. $\dfrac{2}{125}x^2 = \displaystyle\int_0^1 x^2 t^2\, u^2(t)\, dt$

7.4 Nonlinear Weakly-Singular Fredholm Integral Equations

In Chapter 6, we studied the linear weakly-singular Fredholm integral equations of the second kind where the modified decomposition method was applied effectively. In this section we will study the nonlinear weakly-singular

Fredholm integral equations of the second kind. The nonlinear weakly-singular Fredholm integral equations are of the form

$$u(x) = f(x) + \int_0^1 \frac{1}{\sqrt{|x - t|}} F(u(t))dt, \ x \in [0,1], \qquad (144)$$

and its generalized form is given by

$$u(x) = f(x) + \int_0^1 \frac{1}{[|g(x) - g(t)|]^{\alpha}} F(u(t))dt, 0 < \alpha < 1, x \in [0,1], \quad (145)$$

where $F(u(t))$ is a nonlinear function of $u(t)$, such as $u^2(x), u^3(x), e^{u(x)}$, etc.

The nonlinear weakly-singular Fredholm integral equations often arise, as in the linear case, in practical applications such as Dirichlet problems, radiative equilibrium models, potential theory, electrostatics problems, the particle transport problems of astrophysics and reactor theory. Note that that the function $f(x)$ is a given real-valued function. Recall that the kernel is called weakly-singular because singularity can be removed by an appropriate transformation.

Many analytical and numerical methods have been used for the determination of exact or numerical approximations of the solutions of the nonlinear weakly-singular Fredholm integral equations. Examples of the methods that were used so far are the multi-projection method and its re-iterated algorithm, the fast spectral method, the Galerkin collocation method, quadrature methods, the homotopy analysis method, the Taylor series method, the Adomian decomposition method, and the variational iteration method.

In this section, we will use the modified Adomian decomposition method (ADM) and the noise terms phenomenon to determine exact solutions for nonlinear weakly-singular Fredholm integral equations. The modified method was used thoroughly in this text. However, in what follows, we briefly summarize the necessary steps of this method.

7.4.1 The Modified Decomposition Method

The Adomian decomposition method, well-known now in the literature, decomposes the solution of any equation as an infinite series of components, where these components are determined by a recurrence relation. However, the modified decomposition method introduces a slight change in the recurrence relation suggested by the Adomian method. The modified method decomposes the data function $f(x)$ into two components $f_0(x)$ and $f_1(x)$, where only $f_0(x)$ is assigned to the zeroth solution component $u_0(x)$, and the $f_1(x)$ is added to the first component $u_1(x)$ in addition to the other terms assigned by using the standard ADM.

In other words, the modified method proposes the modified recurrence relation given as

$$
\begin{aligned}
u_0(x) &= f_0(x), \\
u_1(x) &= f_1(x) + \int_a^b K(x,t)A_0(t)\,dt, \\
u_{n+1}(x) &= \int_a^b K(x,t)A_n(t)\,dt, n \geq 1.
\end{aligned}
\tag{146}
$$

The use of the modified decomposition method not only minimizes the computations, but avoids the use of the higher order Adomian polynomials. It is worth noting that a proper selection of $f_0(x)$ and $f_1(x)$ is essential for a successful use of the modified decomposition method. However, a criteria for this selection was not found, and trial is the only option.

Recall that we can facilitate the convergence of the solution by using the noise terms phenomenon that we used before in this text. The noise terms are defined as the identical terms with opposite signs. In both instances, the size of the calculations will be minimized, which validates the efficiency and reliability of the modified method and the noise terms phenomenon.

In what follows we will study three illustrative examples to highlight the use of the modified decomposition method and the noise terms phenomenon. We also aim to confirm the power of this method in handling the nonlinear weakly-singular Fredholm integral equations.

Example 1. Consider the nonlinear weakly-singular Fredholm integral equation

$$
u(x) = x - \frac{2}{3(x\sqrt{x} + \sqrt{|x^3 - 1|})} + \int_0^1 \frac{u^2(t)}{\sqrt{|x^3 - t^3|}}\,dt, 0 \leq x \leq 1. \tag{147}
$$

By using the modified decomposition method, the data function $f(x) = x - \frac{2}{3(x\sqrt{x}+\sqrt{|x^3-1|})}$ is decomposed into two parts as

$$
\begin{aligned}
f_0(x) &= x, \\
f_1(x) &= -\frac{2}{3(x\sqrt{x} + \sqrt{|x^3 - 1|})}.
\end{aligned}
\tag{148}
$$

Consequently, we set the modified recurrence relation as

$$
\begin{aligned}
u_0(x) &= x, \\
u_1(x) &= -\frac{2}{3(x\sqrt{x} + \sqrt{|x^3 - 1|})} + \int_0^1 \frac{A_0(t)}{\sqrt{|x^3 - t^3|}}\,dt = 0,
\end{aligned}
\tag{149}
$$

where the zeroth-order Adomian polynomial $A_0(x) = u_0^2(x)$. The other components $u_r(x) = 0$ for $r \geq 2$ vanish in the limit. The exact solution,

$$u(x) = x, \tag{150}$$

follows immediately.

However, to use the Adomian decomposition method combined with the noise terms phenomenon, we set the standard recurrence relation

$$u_0(x) = x - \frac{2}{3(x\sqrt{x} + \sqrt{|x^3 - 1|})},$$
$$u_1(x) = \int_0^1 \frac{A_0(t)(t)}{\sqrt{|x^3 - t^3|}} \, dt = \frac{2}{3(x\sqrt{x} + \sqrt{|x^3 - 1|})} + \text{other terms}. \tag{151}$$

The noise terms $\mp \frac{2}{3(x\sqrt{x} + \sqrt{|x^3 - 1|})}$ appear in both $u_0(x)$ and $u_1(x)$. By canceling the noise term from $u_0(x)$ and verifying that the remaining non-canceled term in $u_0(x)$ identically satisfies the original equation (147), the exact solution is therefore given as

$$u(x) = x. \tag{152}$$

Example 2. Consider the nonlinear weakly-singular Fredholm integral equation

$$u(x) = 1 + x - 2\sqrt{x}\left(1 + 2x + \frac{8}{5}x^2 + \frac{16}{35}x^3\right) \tag{153}$$
$$+ \frac{16}{35}\sqrt{x - 1}\left(12 + 13x + 8x^2 + 2x^3\right) + \int_0^1 \frac{u^3(t)}{\sqrt{|x - t|}} \, dt, 0 \leq x \leq 1.$$

We first decompose $f(x)$ into two parts by

$$\begin{aligned} f_0(x) &= 1 + x, \\ f_1(x) &= -2\sqrt{x}\left(1 + 2x + \frac{8}{5}x^2 + \frac{16}{35}x^3\right) \\ &+ \frac{16}{35}\sqrt{x - 1}\left(12 + 13x + 8x^2 + 2x^3\right). \end{aligned} \tag{154}$$

Consequently, the modified recurrence relation is given by

$$\begin{aligned} u_0(x) &= 1 + x, \\ u_1(x) &= -2\sqrt{x}\left(1 + 2x + \frac{8}{5}x^2 + \frac{16}{35}x^3\right) + \frac{1}{10}\int_0^1 \frac{A_0(t)}{\sqrt{|x - t|}} \, dt = 0, \end{aligned} \tag{155}$$

where the zeroth-order Adomian polynomial $A_0(x) = u_0^2(x)$. The exact solution is thus given by

$$u(x) = 1 + x. \tag{156}$$

Example 3. Consider the nonlinear weakly-singular Fredholm integral equation

$$u(x) = \sqrt{\cos x} + 2\sqrt{\sin|x - 1|} - 2\sqrt{\sin|x|} + \int_0^{\frac{\pi}{2}} \frac{u^2(t)}{\sqrt{|\sin x - \sin t|}} dt, \tag{157}$$

where $0 \le x \le \frac{\pi}{2}$. We first decompose $f(x)$ into two parts given as

$$
\begin{aligned}
f_0(x) &= \sqrt{\cos x}, \\
f_1(x) &= 2\sqrt{\sin|x - 1|} - 2\sqrt{\sin|x|}.
\end{aligned} \tag{158}
$$

We next set the modified recurrence relation as

$$
\begin{aligned}
u_0(x) &= \sqrt{\cos x}, \\
u_1(x) &= 2\sqrt{\sin|x - 1|} - 2\sqrt{\sin|x|} + \int_0^{\frac{\pi}{2}} \frac{A_0(t)}{\sqrt{|\sin x - \sin t|}} dt = 0.
\end{aligned} \tag{159}
$$

This in turn gives the exact solution as

$$u(x) = \sqrt{\cos x}. \tag{160}$$

Example 4. Solve the nonlinear weakly-singular Fredholm integral equation

$$u(x) = 2 + 3x + (8 + 24x + 18x^2)(\sqrt{x - 1} - \sqrt{x}) + \int_0^1 \frac{u^2(t)}{\sqrt{|x - t|}} dt, 0 \le x \le 1. \tag{161}$$

We first decompose $f(x)$ into two parts by

$$
\begin{aligned}
f_0(x) &= 2 + 3x, \\
f_1(x) &= (8 + 24x + 18x^2)(\sqrt{x - 1} - \sqrt{x}).
\end{aligned} \tag{162}
$$

Consequently, the modified recurrence relation is given by

$$
\begin{aligned}
u_0(x) &= 2 + 3x, \\
u_1(x) &= (8 + 24x + 18x^2)(\sqrt{x - 1} - \sqrt{x}) + \int_0^1 \frac{A_0(t)}{\sqrt{|x - t|}} dt = 0.
\end{aligned} \tag{163}
$$

The exact solution is thus given by

$$u(x) = 2 + 3x. \tag{164}$$

Exercises 7.4

Solve the nonlinear weakly-singular Fredholm integral equations

1. $u(x) = x - \frac{16}{15}x^{\frac{5}{2}} + \frac{2}{5}\sqrt{x-1}(1 + \frac{4}{3}x + \frac{8}{3}x^2) + \int_0^1 \frac{u^2(t)}{\sqrt{|x-t|}} dt, 0 \le x \le 1.$

2. $u(x) = x - \frac{32}{35}x^{\frac{7}{2}} + \frac{2}{7}\sqrt{x-1}(1 + 6x + 8x^2 + 16x^3) + \int_0^1 \frac{u^3(t)}{\sqrt{|x-t|}} dt, 0 \le x \le 1.$

3. $u(x) = 1 - x - 2\sqrt{x}(1 - \frac{4}{3}x + \frac{8}{15}x^2) + \frac{16}{15}\sqrt{x-1}(1 - 2x + x^2) + \int_0^1 \frac{u^2(t)}{\sqrt{|x-t|}} dt,$
   $0 \le x \le 1.$

4. $u(x) = 1 + x - \frac{3}{2}x^{\frac{2}{3}}(1 + \frac{6}{5}x + \frac{9}{20}x^2) + \frac{1}{40}(x-1)^{\frac{2}{3}}(123 + 90x + 27x^2) +$
   $\int_0^1 \frac{u^2(t)}{(|x-t|)^{\frac{1}{3}}} dt, 0 \le x \le 1.$

5. $u(x) = \sqrt[4]{\cos x} + \frac{3}{2}(\sin|x-1|)^{\frac{2}{3}} - \frac{3}{2}(\sin|x|)^{\frac{2}{3}} + \int_0^{\frac{\pi}{2}} \frac{u^4(t)}{(|\sin x - \sin t|)^{\frac{1}{3}}} dt,$
   $0 \le x \le \frac{\pi}{2}$

6. $u(x) = \sqrt{\sin x} + \frac{4}{3}(\cos|x-1|)^{\frac{3}{4}} - \frac{4}{3}(\cos|x|)^{\frac{3}{4}} + \int_0^{\frac{\pi}{2}} \frac{u^2(t)}{(|\cos x - \cos t|)^{\frac{1}{4}}} dt,$
   $0 \le x \le \frac{\pi}{2}$

Chapter 8

Nonlinear Volterra Integral Equations

8.1 Introduction

In the previous chapter, we studied the nonlinear Fredholm integral equations of the first and the second kinds. We pointed out earlier that nonlinear integral equations need a considerable amount of work. However, with the recent developed methods we can minimize significantly this cumbersome work. It is therefore useful to present some reliable and powerful techniques that will make the study of nonlinear integral equations successful and valuable. In general, the solution of the nonlinear integral equations is not in general unique as we studied in the case of the nonlinear Fredholm integral equations. However, the existence of a unique solution of nonlinear integral equations with specific conditions is possible but cannot be assumed general. This will be illustrated by the forthcoming examples. Accordingly, our emphasis will be on introducing reliable and easily calculable techniques for solving specific cases of nonlinear Volterra integral equations. As indicated in Chapter 1, integral equations of the form

$$u(x) = f(x) + \lambda \int_a^b K(x,t) \, F(u(t)) dt \tag{1}$$

and

$$u(x) = f(x) + \lambda \int_0^x K(x,t) \, F(u(t)) dt, \tag{2}$$

are called *nonlinear* Fredholm integral equations and *nonlinear* Volterra integral equations respectively. The function $F(u(x))$ is nonlinear in $u(x)$

such as $u^2(x), u^3(x), e^{u(x)}, \sin u(x)$ and many others, and λ is a parameter. In this text, we will concern ourselves to the case where $F(u(t)) = u^n(t)$, $n \geq 2$, whereas other nonlinear integral equations that involve other forms of nonlinearity of $F(u(x))$ can be handled in a very similar way. The following are examples of the *nonlinear* Volterra integral equations of the second kind

$$u(x) = x - \frac{1}{4}x^4 + \int_0^x tu^2(t)dt, \tag{3}$$

$$u(x) = 2x + \frac{1}{6}x^5 - \int_0^x tu^3(t)\,dt. \tag{4}$$

Moreover, the nonlinear Volterra integral equations of the first kind are of the form

$$x = \int_0^x tu^2(t)dt, \tag{5}$$

$$2x = -\int_0^x t^2u^4(t)\,dt. \tag{6}$$

8.2   Nonlinear Volterra Integral Equations of the Second Kind

In this section we will focus our study on the nonlinear Volterra integral equations of the second kind given in the standard form

$$u(x) = f(x) + \lambda \int_0^x K(x,t)\,F(u(t))dt, \tag{7}$$

where $F(u(t))$ is nonlinear in $u(t)$ and λ is a parameter. Several powerful methods, analytical and numerical, are used in the literature. Based on our discussions in Chapters 3 and 5 we have found that the *series solution method*, the *Adomian decomposition method*, and the *variational iteration method* proved to be reliable techniques in handling successfully the linear Volterra integral equations and the linear Volterra integro-differential equations. It seems reasonable to use these methods in our study of the nonlinear Volterra integral equations. It is to be noted that the Adomian decomposition method approaches the nonlinear Volterra equations generally by using the so called Adomian polynomials, that was introduced earlier in Chapter 7, hence we will skip details. The so-called Adomian polynomials will be used to represent the involved nonlinear function $u^n(t), n \geq 2$ in a similar manner as presented before. In the following we will outline the steps needed to use these methods effectively.

8.2.1   The Series Solution Method

The nonlinear Volterra integral equations of the form

$$u(x) = f(x) + \lambda \int_0^x K(x,t)\, u^n(t)dt,\qquad (8)$$

where the kernel $K(x,t)$ will be assumed a separable kernel, will be examined using the *series solution method*. To use this method we should assume that $u(x)$ is analytic, hence it admits the Taylor expansion about $x = 0$ given by

$$u(x) = \sum_{n=0}^{\infty} a_n x^n.\qquad (9)$$

Substituting (9) into both sides of (8), assuming that $K(x,t) = h(x)g(t)$ yields

$$\sum_{n=0}^{\infty} a_n x^n = f(x) + \lambda h(x) \int_0^x g(t) \left(\sum_{n=0}^{\infty} a_n t^n \right)^n dt,\qquad (10)$$

or simply

$$a_0 + a_1 x + a_2 x^2 + \cdots = f(x) + \lambda h(x) \int_0^x g(t) \left(a_0 + a_1 t + a_2 t^2 + \cdots \right)^n dt,\qquad (11)$$

such that the integral in (8) that includes the unknown function $u(x)$ is reduced to an easily computable integrals. Using the Taylor expansions for $f(x)$ and $h(x)$, integrating the resulting simple integral at the right hand side, and then equating the coefficients of like powers of x lead to the complete determination of the coefficients $a_i, i \geq 0$. Consequently, the solution $u(x)$ is readily obtained upon using (9). As discussed before, the exact solution may be obtained if the resulting series is an expansion of a well known function, otherwise we use few terms of the obtained series to achieve an accurate numerical approximation for computational purposes. The method discussed above will be illustrated by using the following examples.

Example 1. We first consider the nonlinear Volterra integral equation

$$u(x) = x - \frac{1}{4}x^4 + \int_0^x t u^2(t)\, dt.\qquad (12)$$

Substituting the series form of $u(x)$ given by (9) into both sides of (12) yields

$$a_0 + a_1 x + a_2 x^2 + a_3 x^3 + \cdots$$
$$= x - \tfrac{1}{4}x^4 + \int_0^x t \left[a_0 + a_1 t + a_2 t^2 + a_3 t^3 + \cdots \right]^2 dt,\qquad (13)$$

or equivalently

$$a_0 + a_1 x + a_2 x^2 + a_3 x^3 + \cdots$$

$$= x - \tfrac{1}{4}x^4 + \int_0^x t \left[a_0^2 + 2a_0 a_1 t + \left(2a_0 a_2 + a_1^2 \right) t^2 + \cdots \right] dt. \tag{14}$$

Integrating the integral at the right hand side of (14) we find

$$a_0 + a_1 x + a_2 x^2 + a_3 x^3 + \cdots = x - \tfrac{1}{4}x^4$$

$$+ \frac{1}{2}a_0^2 x^2 + \frac{2}{3}a_0 a_1 x^3 + \frac{1}{4}\left(a_1^2 + 2a_0 a_2 \right) x^4 + \cdots . \tag{15}$$

Equating the coefficients of like powers of x in both sides yields

$$a_1 = 1, \quad a_n = 0, \quad \text{for} \quad n \neq 1. \tag{16}$$

Consequently, the exact solution is given by

$$u(x) = x, \tag{17}$$

upon substituting (16) into (9).

Example 2. We next consider the nonlinear Volterra integral equation

$$u(x) = e^x - \frac{1}{3}xe^{3x} + \frac{1}{3}x + \int_0^x xu^3(t)dt. \tag{18}$$

Substituting the series (9) into both sides of (18) noting that

$$u^3(x) = a_0^3 + 3a_0^2 a_1 x + (3a_0 a_1^2 + 3a_0^2 a_2)x^2 + \cdots \tag{19}$$

gives

$$a_0 + a_1 x + a_2 x^2 + a_3 x^3 + a_4 x^4 + \cdots$$

$$= \left(1 + x + \frac{x^2}{2!} + \frac{x^3}{3!} + \frac{x^4}{4!} + \cdots \right)$$

$$- \frac{1}{3}x \left(1 + 3x + \frac{9}{2}x^2 + \frac{27}{3!}x^3 + \frac{81}{4!}x^4 + \cdots \right) + \frac{1}{3}x \tag{20}$$

$$+ x \int_0^x \left[a_0^3 + 3a_0^2 a_1 t + (3a_0 a_1^2 + 3a_0^2 a_2)t^2 + \cdots \right] dt.$$

Evaluating the integral at the right hand side of (20) and equating the coefficients of like powers of x we find

$$\begin{aligned}
a_0 &= 1 \\
a_1 &= 1 \\
a_2 &= \frac{1}{2!} \\
a_3 &= \frac{1}{3!}
\end{aligned}$$

(21)

$$\vdots$$

$$a_n = \frac{1}{n!} \quad \text{for } n \geq 0.$$

This gives the solution in a series form

$$u(x) = 1 + x + \frac{1}{2!}x^2 + \frac{1}{3!}x^3 + \cdots$$

(22)

which gives the solution in a closed form by

$$u(x) = e^x.$$

(23)

Example 3. We now consider the nonlinear Volterra integral equation

$$u(x) = \cos x - \frac{1}{4}\sin(2x) - \frac{1}{2}x + \int_0^x u^2(t)dt.$$

(24)

Note that we should use the Taylor series of $\cos x$ and $\sin(2x)$ by

$$\begin{aligned}
\cos x &= 1 - \frac{1}{2!}x^2 + \frac{1}{4!}x^4 + \cdots, \\
\sin(2x) &= (2x) - \frac{1}{3!}(2x)^3 + \frac{1}{5!}(2x)^5 + \cdots.
\end{aligned}$$

(25)

Substituting the series (9) into both sides of (24) gives

$$\begin{aligned}
&a_0 + a_1 x + a_2 x^2 + a_3 x^3 + a_4 x^4 + \cdots \\
&= 1 - x - \frac{1}{2}x^2 + \frac{1}{3}x^3 + \frac{1}{24}x^4 + \cdots \\
&+ \int_0^x \left[a_0 + a_1 t + a_2 t^2 + \cdots \right]^2 dt.
\end{aligned}$$

(26)

Evaluating the integral at the right hand side of (26) and equating the coefficients of like powers of x we find

$$\begin{aligned}
a_0 &= 1 \\
a_1 &= 0 \\
a_2 &= \frac{1}{2!} \\
a_3 &= 0 \\
a_4 &= \frac{1}{4!},
\end{aligned}$$

(27)

$$\vdots$$

This gives the solution in a series form

$$u(x) = 1 - \frac{1}{2!}x^2 + \frac{1}{4!}x^4 + \cdots,\tag{28}$$

which gives the solution in a closed form by

$$u(x) = \cos x.\tag{29}$$

Exercises 8.2.1

Use the *series solution method* to solve the following nonlinear Volterra integral equations:

1. $u(x) = x^2 + \frac{1}{10}x^5 - \frac{1}{2}\int_0^x u^2(t)\,dt.$

2. $u(x) = x^2 + \frac{1}{12}x^6 - \frac{1}{2}\int_0^x tu^2(t)\,dt.$

3. $u(x) = 1 - x^2 - \frac{1}{3}x^3 + \int_0^x u^2(t)\,dt.$

4. $u(x) = 1 - x + x^2 - \frac{2}{3}x^3 - \frac{1}{5}x^5 + \int_0^x u^2(t)\,dt.$

5. $u(x) = x^2 + \frac{1}{14}x^7 - \frac{1}{2}\int_0^x u^3(t)\,dt.$

6. $u(x) = \frac{1}{2} + e^{-x} - \frac{1}{2}e^{-2x} - \int_0^x u^2(t)\,dt.$

7. $u(x) = 1 - \frac{3}{2}x^2 - x^3 - \frac{1}{4}x^4 + \int_0^x u^3(t)\,dt.$

8. $u(x) = \sin x - \frac{1}{2}x + \frac{1}{4}\sin 2x + \int_0^x u^2(t)\,dt.$

9. $u(x) = \cos x - \frac{1}{2}x - \frac{1}{4}\sin 2x + \int_0^x u^2(t)\,dt,$

10. $u(x) = e^x + \frac{1}{2}x(e^{2x} - 1) - \int_0^x xu^2(t)\,dt.$

8.2.2   The Adomian Decomposition Method

The purpose of this section is to describe how the Adomian decomposition method can be applied to nonlinear Volterra integral equations. Even though the method does not discuss the existence and the uniqueness concepts, but it provides a reliable and powerful technique to handle nonlinear equations.

The method has been introduced in details in solving nonlinear Fredholm integral equations. For this reason, we will outline a brief framework of the method that will be implemented for nonlinear Volterra integral equations. Recall that we will focus our study on the nonlinear Volterra integral equations of the form

$$u(x) = f(x) + \lambda \int_0^x K(x,t)\, u^n(t) dt, \tag{30}$$

where the kernel is assumed a separable kernel. We usually represent the solution $u(t)$ of (30) by the series

$$u(x) = \sum_{n=0}^{\infty} u_n(x), \tag{31}$$

and the nonlinear term $u^n(t)$, under the integral sign of the equation (30), by the polynomial series

$$u^n(t) = \sum_{n=0}^{\infty} A_n(t), \tag{32}$$

where the $A_n(t)\, n \geq 0$ are the so called Adomian polynomials. The Adomian polynomials can be established by using the algorithm

$$\begin{cases} A_0 = F(u_0), \\ A_1 = u_1 \dfrac{d}{du_0} F(u_0), \\ A_2 = u_2 \dfrac{d}{du_0} F(u_0) + \dfrac{u_1^2}{2!} \dfrac{d^2}{du_0^2} F(u_0), \\ A_3 = u_3 \dfrac{d}{du_0} F(u_0) + u_1 u_2 \dfrac{d^2}{du_0^2} F(u_0) + \dfrac{u_1^3}{3!} \dfrac{d^3}{du_0^3} F(u_0), \\ \vdots \end{cases} \tag{33}$$

We introduced in the previous section several examples explaining how we can generate Adomian polynomials.

Substituting (31) and (32) into (30) yields

$$\sum_{n=0}^{\infty} u_n(x) = f(x) + \lambda \int_0^x K(x,t) \left(\sum_{n=0}^{\infty} A_n(t) \right) dt \tag{34}$$

or simply

$$\begin{aligned} u_0(x) + u_1(x) + u_2(x) + \cdots &= f(x) \\ + \lambda \int_0^x K(x,t) \left[A_0(t) + A_1(t) + A_2(t) + \cdots \right] dt. \end{aligned} \tag{35}$$

The components $u_0(x), u_1(x), u_2(x), \cdots$ are completely determined by using the recurrent scheme

$$
\begin{cases}
u_0(x) \ = \ f(x), \\
u_1(x) \ = \ \lambda \displaystyle\int_0^x K(x,t)A_0(t)dt, \\
u_2(x) \ = \ \lambda \displaystyle\int_0^x K(x,t)A_1(t)dt, \\
\vdots \\
u_{n+1}(x) = \lambda \displaystyle\int_a^b K(x,t)A_n(t)dt, \quad n \geq 0.
\end{cases}
\tag{36}
$$

Consequently the solution of (30) in a series form follows immediately by using (31). As indicated earlier, the series obtained may yield the exact solution in a closed form, or a truncated series $\sum_{n=1}^{k} u_n(x)$ may be used if a numerical approximation is desired.

It is worth noting that the *modified decomposition method* and the *noise terms phenomenon* work also effectively for nonlinear cases and play a major role in minimizing the size of calculations. In what follows, we will apply the Adomian decomposition method and the modified decomposition method for handling the nonlinear Volterra integral equations.

Example 1. We first consider the nonlinear Volterra integral equation

$$
u(x) = x + \frac{1}{5}x^5 - \int_0^x tu^3(t)\, dt.
\tag{37}
$$

We start by setting the zeroth component

$$
u_0(x) = x + \frac{1}{5}x^5,
\tag{38}
$$

so that the first component is obtained by

$$
u_1(x) = - \int_0^x tA_0(t)\, dt,
\tag{39}
$$

which gives

$$
u_1(x) = -\frac{1}{5}x^5 - \frac{1}{15}x^9 - \frac{3}{325}x^{13} - \frac{1}{2125}x^{17},
\tag{40}
$$

upon using

$$
A_0(t) = (t + \frac{1}{5}t^5)^3.
\tag{41}
$$

It can be easily observed that by cancelling the noise terms $\frac{1}{5}x^5$ and $-\frac{1}{5}x^5$ between $u_0(x)$ and $u_1(x)$, and justifying that the remaining term of $u_0(x)$ justifies the equation lead to the exact solution

$$u(x) = x. \tag{42}$$

We point out here that the exact solution can also be obtained by using the modified decomposition method.

Example 2. We next consider the nonlinear Volterra integral equation

$$u(x) = 2x - \frac{1}{12}x^4 + \frac{1}{4}\int_0^x (x - t)u^2(t)\, dt. \tag{43}$$

To minimize the calculations volume, we will use the modified decomposition method in this example. For this reason we split $f(x)$ between the two components $u_0(x)$ and $u_1(x)$, hence we set

$$u_0(x) = 2x. \tag{44}$$

Consequently, the first component is defined by

$$u_1(x) = -\frac{1}{12}x^4 + \frac{1}{4}\int_0^x (x - t)A_0(t)\, dt, \tag{45}$$

which gives

$$u_1(x) = 0, \tag{46}$$

upon using

$$A_0(t) = (2t)^2. \tag{47}$$

This defines the other components by

$$u_k(x) = 0, \quad \text{for } k \geq 1. \tag{48}$$

The exact solution

$$u(x) = 2x, \tag{49}$$

follows immediately.

Example 3. It seems reasonable to compare the *series solution method* and the *Adomian decomposition method* by solving Example 2 in the previous subsection given by

$$u(x) = e^x - \frac{1}{3}xe^{3x} + \frac{1}{3}x + \int_0^x xu^3(t)dt. \tag{50}$$

Applying the standard decomposition method will result in a considerable amount of difficulties in integrating and forming the Adomian polynomials. It is useful to consider using the *modified decomposition method*. Splitting $f(x)$ between the first two components yields

$$u_0(x) = e^x, \tag{51}$$

and

$$u_1(x) = -\frac{1}{3}xe^{3x} + \frac{1}{3}x + \int_0^x xA_0(t)dt, \tag{52}$$

or equivalently

$$u_1(x) = 0 \tag{53}$$

upon using

$$A_0(t) = (e^{3t}). \tag{54}$$

Accordingly, the exact solution

$$u(x) = e^x, \tag{55}$$

is readily obtained.

Exercises 8.2.2

Use the *decomposition method* or the *modified decomposition method* to solve the following nonlinear Volterra integral equations by finding the exact solution or by writing few terms of the series solution

1. $u(x) = 3x + \dfrac{1}{24}x^4 - \dfrac{1}{18}\displaystyle\int_0^x (x-t)u^2(t)\, dt.$

2. $u(x) = 2x - \dfrac{1}{2}x^4 + \dfrac{1}{4}\displaystyle\int_0^x u^3(t)\, dt.$

3. $u(x) = \sin x + \dfrac{1}{8}\sin(2x) - \dfrac{1}{4}x + \dfrac{1}{2}\displaystyle\int_0^x u^2(t)\, dt.$

4. $u(x) = x^2 + \dfrac{1}{5}x^5 - \displaystyle\int_0^x u^2(t)\, dt.$

5. $u(x) = x + \displaystyle\int_0^x (x-t)u^3(t)\, dt.$

6. $u(x) = 1 + \displaystyle\int_0^x (x-t)^2 u^2(t)\, dt.$

7. $u(x) = 1 + \displaystyle\int_0^x (x-t)^2 u^3(t)\, dt.$

8. $u(x) = x + \displaystyle\int_0^x (x-t)^2 u^2(t)\, dt.$

9. $u(x) = 1 + \displaystyle\int_0^x \left(t + u^2(t)\right)\, dt.$

10. $u(x) = 1 + \displaystyle\int_0^x \left(t^2 + u^2(t) \right) \, dt.$

11. $u(x) = \sec x + \tan x + x - \displaystyle\int_0^x \left(1 + u^2(t) \right) \, dt, \ x < \dfrac{\pi}{2}.$

12. $u(x) = \tan x - \dfrac{1}{4} \sin(2x) - \dfrac{x}{2} + \displaystyle\int_0^x \dfrac{1}{1 + u^2(t)} \, dt, \ x < \dfrac{\pi}{2}.$

8.2.3   The Variational Iteration Method

The *variational iteration method* was presented in the preceding chapters for handling integral equations. The method has been proved to be reliable in the study of linear and nonlinear, and homogeneous and inhomogeneous equations. The method gives the successive approximations of the exact solution that may converge to the exact solution in case this solution exists. One significant feature of this method is that it can handle especially nonlinear problems without the use of the so-called Adomian polynomials as required by the Adomian decomposition method.

The variational iteration method admits the use of a correction functional in the form

$$u_{n+1}(x) = u_n(x) + \int_0^x \lambda(\xi) \left(L u_n(\xi) + N \, F(u_n(\xi)) - g(\xi) \right) \, d\xi, n \geq 0, \quad (56)$$

where λ is a general Lagrange multiplier that can be determined optimally via the variational theory as shown before and $F(u(x))$ is a nonlinear function of $u(x)$. In Chapter 2, we presented a rule that gives these Lagrange multipliers for some ordinary differential equations that will be examined in this text. Recall that the variational iteration method (VIM) is used for ODEs and integro-differential equations and this can be obtained by differentiating the Volterra integral equation. Moreover, the zeroth component $u_0(x)$ in (56) can be selected according to the order of the resulted ODE. The exact solution is thus given by

$$u(x) = \lim_{n \to \infty} u_n(x). \quad (57)$$

In other words, to solve any nonlinear Volterra integral equation by using the variational method, we should first transform this equation to its equivalent ODE or its equivalent nonlinear Volterra integro-differential equation by differentiating both sides, where Leibniz rule should be used. The next step consists of the determination of the Lagrange multiplier λ using the rules given in Chapter 2. Finally, we select the zeroth approximation $u_0(x)$ as indicated earlier. Having prepared all these steps, we then use the correction functional (56) to determine as many successive approximations as we can. For simplicity reasons, we will focus our study on the difference kernel, where $K(x, t) = K(x - t)$.

Example 1. Solve the nonlinear Volterra integral equation by using the variational iteration method

$$u(x) = 2x - \frac{1}{12}x^4 + \frac{1}{4}\int_0^x (x-t)u^2(t)\, dt. \tag{58}$$

Differentiating both sides of this equation, and using Leibniz rule, we find

$$u'(x) = 2 - \frac{1}{3}x^3 + \frac{1}{4}\int_0^x u^2(t)\, dt, u(0) = 0. \tag{59}$$

The initial condition $u(0) = 0$ is obtained by using $x = 0$ into the integral equation, and hence we can select $u_0(x) = 0$.

The correction functional for equation (59) is

$$u_{n+1}(x) = u_n(x) - \int_0^x \left(u_n'(\xi) - 2 + \frac{1}{3}\xi^3 - \frac{1}{4}\int_0^\xi u_n^2(r)\, dr \right)\, d\xi, \tag{60}$$

where we selected $\lambda = -1$ for the first order integro-differential equation. As stated before, we can use the initial condition to select $u_0(x) = 0$ that will lead to the following successive approximations

$$
\begin{aligned}
u_0(x) &= 0, \\
u_1(x) &= 2x - \frac{1}{12}x^4, \\
u_2(x) &= 2x - \frac{1}{12}x^4 + (\frac{1}{12}x^4 - \frac{1}{504}x^7) + \frac{1}{51840}x^{10}, \\
u_3(x) &= 2x - \frac{1}{504}x^7 + (\frac{1}{504}x^7 + \frac{1}{51840}x^{10}) - \frac{1}{51840}x^{10} + \cdots,
\end{aligned} \tag{61}
$$

$$\vdots$$

Cancelling the noise terms gives the exact solution by

$$u(x) = 2x. \tag{62}$$

Example 2. Solve the nonlinear Volterra integral equation by using the variational iteration method

$$u(x) = x^2 - \frac{1}{30}x^6 + \int_0^x (x-t)u^2(t)\, dt. \tag{63}$$

Differentiating both sides of this equation, and using Leibniz rule, we find

$$u'(x) = 2x - \frac{1}{5}x^5 + \int_0^x u^2(t)\, dt, u(0) = 0. \tag{64}$$

The initial condition $u(0) = 0$ is obtained by using $x = 0$ into the integral equation, and hence we can select $u_0(x) = 0$.

The correction functional for equation (64) is

$$u_{n+1}(x) = u_n(x) - \int_0^x \left(u_n'(\xi) - 2\xi + \frac{1}{5}\xi^5 - \int_0^\xi u_n^2(r)\, dr \right) d\xi, \quad (65)$$

where we selected $\lambda = -1$ for the first order integro-differential equation. As stated before, we can use the initial condition to select $u_0(x) = 0$ that will lead to the following successive approximations

$$
\begin{aligned}
u_0(x) &= 0, \\
u_1(x) &= x^2 - \tfrac{1}{30}x^6, \\
u_2(x) &= x^2 - \tfrac{1}{30}x^6 + (\tfrac{1}{30}x^6 - \tfrac{1}{1350}x^{10}) + \tfrac{1}{163800}x^{14}, \\
u_3(x) &= x^2 - \tfrac{1}{1350}x^{10} + (\tfrac{1}{1350}x^{10} + \tfrac{1}{163800}x^{14}) - \tfrac{1}{163800}x^{14} + \cdots, \\
&\vdots
\end{aligned}
$$

$$(66)$$

Cancelling the noise terms gives the exact solution by

$$u(x) = x^2. \quad (67)$$

Example 3. Solve the nonlinear Volterra integral equation by using the variational iteration method

$$u(x) = x + x^3 - \frac{1}{30}x^5 - \frac{2}{105}x^7 - \frac{1}{252}x^9 + \int_0^x (x-t)^2 u^2(t)\, dt. \quad (68)$$

Differentiating both sides of this equation, and using Leibniz rule, we find

$$u'(x) = 1 + 3x^2 - \frac{1}{6}x^4 - \frac{2}{15}x^6 - \frac{1}{28}x^8 + 2\int_0^x (x-t)u^2(t)\, dt,\ u(0) = 0. \quad (69)$$

The initial condition $u(0) = 0$ is obtained by using $x = 0$ into the integral equation, and hence we can select $u_0(x) = 0$. the following successive approximations

$$
\begin{aligned}
u_0(x) &= 0, \\
u_1(x) &= x + x^3 - \tfrac{1}{30}x^5 - \tfrac{2}{105}x^7 - \tfrac{1}{252}x^9, \\
u_2(x) &= x + x^3 - \tfrac{1}{3780}x^9 + \cdots, \\
u_3(x) &= x + x^3 + \cdots, \\
&\vdots
\end{aligned}
$$

$$(70)$$

Cancelling the noise terms gives the exact solution by

$$u(x) = x + x^3. \quad (71)$$

Exercises 8.2.3

Use the *variational iteration method* to solve the following nonlinear Volterra integral equations

1. $u(x) = x - \dfrac{1}{20}x^5 + \displaystyle\int_0^x (x - t)u^3(t)\,dt.$

2. $u(x) = x^2 - \dfrac{1}{56}x^8 + \displaystyle\int_0^x (x - t)u^3(t)\,dt.$

3. $u(x) = x + x^2 - \dfrac{1}{12}x^4 - \dfrac{1}{10}x^5 - \dfrac{1}{30}x^6 + \displaystyle\int_0^x (x - t)u^2(t)\,dt.$

4. $u(x) = 1 + x - \dfrac{1}{2}x^2 - \dfrac{1}{3}x^3 - \dfrac{1}{12}x^4 + \displaystyle\int_0^x (x - t)u^2(t)\,dt.$

5. $u(x) = 1 + x - \dfrac{1}{3}x^3 - \dfrac{1}{4}x^4 - \dfrac{1}{30}x^5 + \displaystyle\int_0^x (x - t)^2 u^2(t)\,dt.$

8.3   Nonlinear Volterra Integral Equations of the First Kind

In this section we will study the nonlinear Volterra integral equation of the first kind that reads

$$f(x) = \int_0^x K(x,t)F(u(t))\,dt, \tag{72}$$

where the kernel $K(x,t)$ and the function $f(x)$ are given real-valued functions. The function $F(u(x))$ is a nonlinear function of $u(x)$. As a first kind equation, the unknown function $u(x)$ appears only under the integral sign, and as a Volterra equation there is at least one limit of integration is a variable x. The linear Volterra integral equation of the first kind is presented in Section 3.9 where it was examined by using a variety of powerful methods.

In this section we will first use the series solution method. However, it is to be noted that this method suffers from the size of computational work, that can be facilitated by using any computer algebra system such as Maple or Mathematica. The series method can be used in a direct manner like the way presented earlier in this Chapter.

8.3.1   The Series Solution Method

The series solution method was presented earlier in details. The nonlinear Volterra integral equations of the first kind reads

$$f(x) = \int_0^x K(x,t)\,u^n(t)\,dt, \tag{73}$$

where the kernel $K(x,t)$ will be assumed a separable kernel, will be examined using the *series solution method*. For simplicity, we used the nonlinear term $u^n(x)$ for the nonlinear general operator $F(u(x))$. To use this method we should assume that $u(x)$ is given by

$$u(x) = \sum_{n=0}^{\infty} a_n x^n. \tag{74}$$

Substituting (74) into both sides of (73), assuming that $K(x,t) = h(x)g(t)$ yields

$$f(x) = h(x) \int_0^x g(t) \left(\sum_{n=0}^{\infty} a_n t^n \right)^n dt, \tag{75}$$

or simply

$$f(x) = h(x) \int_0^x g(t) \left(a_0 + a_1 t + a_2 t^2 + \cdots \right)^n dt. \tag{76}$$

Using the Taylor expansions for $f(x)$ and $h(x)$, integrating the resulting integral at the right side, and then equating the coefficients of like powers of x lead to the complete determination of the coefficients $a_i, i \geq 0$. Consequently, the solution $u(x)$ is readily obtained upon using (74). As stated earlier, the method requires more work compared to the newly developed methods such as the Adomian decomposition method or the variational iteration method. In what follows we will use the series method to explain three nonlinear Volterra integral equations of the first kind.

Example 1. We first consider the nonlinear Volterra integral equation of the first kind

$$\frac{1}{2}e^{2x} - \frac{1}{2} = \int_0^x u^2(t) dt. \tag{77}$$

Using the Taylor series of the left side, and substituting the series form of $u(x)$ given by (74) into both sides of (77) yields

$$x + x^2 + \frac{2}{3}x^3 + \frac{1}{3}x^4 + \cdots = \int_0^x \left[a_0 + a_1 t + a_2 t^2 + a_3 t^3 + \cdots \right]^2 dt. \tag{78}$$

Integrating the integral at the right side of (78) we find

$$x + x^2 + \frac{2}{3}x^3 + \frac{1}{3}x^4 + \cdots = a_0^2 x + a_0 a_1 x^2 + \frac{1}{3}(2a_0 a_2 + a_1^2)x^3 + \cdots. \tag{79}$$

Equating the coefficients of like powers of x in both sides yields

$$a_0 = \pm 1, \quad a_1 = \pm 1, \quad a_2 = \pm \frac{1}{2!}, \quad a_3 = \pm \frac{1}{3!}, \cdots \tag{80}$$

Consequently, the exact solution is given by

$$u(x) = \pm e^x. \tag{81}$$

Notice that we did not use Adomian polynomials in solving this problem. Moreover, the series solution method gave use two answers and this is normal for non linear problem.

Example 2. We next consider the nonlinear Volterra integral equation of the first kind

$$\frac{1}{4}x^2 + \frac{1}{4}\sin^2 x = \int_0^x (x - t)u^2(t)\, dt. \tag{82}$$

Using the Taylor series of the left side, and substituting the series form of $u(x)$ given by (74) into both sides of (82) yields

$$\frac{1}{2}x^2 - \frac{1}{12}x^4 + \frac{1}{90}x^6 + \cdots = \int_0^x (x - t)\left[a_0 + a_1 t + a_2 t^2 + a_3 t^3 + \cdots\right]^2 dt. \tag{83}$$

Proceeding as before, and equating the coefficients of like powers of x in both sides yields

$$a_0 = \pm 1, \quad a_1 = 0, \quad a_2 = \pm\frac{1}{2!}, \quad a_3 = 0, \quad a_4 = \pm\frac{1}{4!}, \cdots \tag{84}$$

Consequently, the exact solution is given by

$$u(x) = \pm \cos x. \tag{85}$$

Example 3. We next consider the nonlinear Volterra integral equation of the first kind

$$\frac{1}{2}x + \frac{1}{2}x^2 - \frac{1}{4}\sin(2x) = \int_0^x (x - t)u^2(t)\, dt. \tag{86}$$

Using the Taylor series of the left side, and substituting the series form of $u(x)$ given by (74) into both sides of (86) yields

$$\frac{1}{2}x^2 - \frac{1}{3}x^3 - \frac{1}{15}x^5 + \cdots = \int_0^x (x-t)\left[a_0 + a_1 t + a_2 t^2 + a_3 t^3 + \cdots\right]^2 dt. \tag{87}$$

Proceeding as before, and equating the coefficients of like powers of x in both sides yields

$$a_0 = \pm 1, \quad a_1 = \pm 1, \quad a_2 = \mp\frac{1}{2!}, \quad a_3 = \mp\frac{1}{3!}, \quad a_4 = \pm\frac{1}{4!}, \cdots \tag{88}$$

Consequently, the exact solution is given by

$$u(x) = \pm(\cos x + \sin x). \tag{89}$$

Exercises 8.3.1

Use the *series solution method* to solve the following nonlinear Volterra integral equations of the first kind:

1. $\dfrac{1}{4}x^2 - \dfrac{1}{4}\sin^2 x = \displaystyle\int_0^x (x - t)u^2(t)\, dt.$

2. $\dfrac{1}{4}e^{2x} - \dfrac{1}{2}x - \dfrac{1}{4} = \displaystyle\int_0^x (x - t)u^2(t)\, dt.$

3. $\dfrac{1}{3}e^{3x} - \dfrac{1}{3} = \displaystyle\int_0^x u^3(t)\, dt.$

4. $\dfrac{1}{4}\sin(2x) + \dfrac{1}{4}x^2 - \dfrac{1}{2}x = \displaystyle\int_0^x (x - t)u^2(t)\, dt.$

8.3.2 Conversion to a Volterra Equation of the Second Kind

We turn now to use the commonly used methods, used in this text, to handle the nonlinear Volterra integral equations of the first kind. To determine a solution for the nonlinear Volterra integral equation of the first kind (72), we follow our approach in the linear case in Section 3.9, hence we first convert it to a linear Volterra integral equation of the first kind of the form

$$f(x) = \int_0^x K(x,t)v(t)\, dt, \tag{90}$$

by using the transformation

$$v(x) = F(u(x)). \tag{91}$$

This in turn means that

$$u(x) = F^{-1}(v(x)). \tag{92}$$

It is worth noting that the Volterra integral equation of the first kind (90) can be solved by any method that was studied in Section 3.9. However, in this section we will handle Eq. (90) by the conversion to Volterra integral equation of the second kind, and this can be achieved by using Leibniz rule. The conversion technique works effectively only if $K(x,x) \neq 0$. Differentiating both sides of (90) with respect to x, and using Leibniz rule, we find

$$f'(x) = K(x,x)v(x) + \int_0^x K_x(x,t)v(t)dt. \tag{93}$$

Solving for $v(x)$, provided that $K(x,x) \neq 0$, we obtain the Volterra integral equation of the second kind given by

$$v(x) = \frac{f'(x)}{K(x,x)} - \int_0^x \frac{1}{K(x,x)} K_x(x,t)v(t)dt. \tag{94}$$

Having converted the Volterra integral equation of the first kind to the Volterra integral equation of the second kind, we then can use any method that was presented before. Because we solved the Volterra integral equations of the second kind by many methods in this text, therefore we will select specific methods for solving the nonlinear Volterra integral equation of the first kind after converting it to a Volterra integral equation of the second kind. This will be worked out for revision purposes only.

Example 1. Convert the nonlinear Volterra integral equation of the first kind to the second kind and solve the resulting equation

$$\frac{1}{5}x^5 + \frac{1}{30}x^6 = \int_0^x (x - t + 1)u^2(t)dt. \tag{95}$$

We first set

$$v(x) = u^2(x), u(x) = \pm\sqrt{v(x)}, \tag{96}$$

to carry out (95) into

$$\frac{1}{5}x^5 + \frac{1}{30}x^6 = \int_0^x (x - t + 1)v(t)dt. \tag{97}$$

Differentiating both sides of (97) with respect to x by using Leibniz rule we find the Volterra integral equation of the second kind

$$v(x) = x^4 + \frac{1}{5}x^5 - \int_0^x v(t)dt. \tag{98}$$

We will select the modified decomposition method to solve this equation. Using the recursive relation

$$\begin{aligned} v_0(x) &= x^4, \\ v_1(x) &= \frac{1}{5}x^5 - \int_0^x v_0(t)dt = 0. \end{aligned} \tag{99}$$

The exact solutions are therefore given by

$$u(x) = \pm x^2. \tag{100}$$

Example 2. Convert the Volterra integral equation of the first kind to the second kind and solve the resulting equation

$$\frac{5}{4}e^{2x} - \frac{1}{2}x - \frac{5}{4} = \int_0^x (x - t + 2)u^2(t)dt. \tag{101}$$

We first set

$$v(x) = u^2(x), u(x) = \pm\sqrt{v(x)}, \tag{102}$$

to carry out (101) into

$$\frac{5}{4}e^{2x} - \frac{1}{2}x - \frac{5}{4} = \int_0^x (x - t + 2)v(t)dt. \tag{103}$$

Differentiating both sides of (103) and proceeding as before we find

$$v(x) = \frac{5}{4}e^{2x} - \frac{1}{4} - \frac{1}{2}\int_0^x v(t)dt. \tag{104}$$

We will select the modified decomposition method to solve this equation. Using the recursive relation

$$\begin{aligned}
v_0(x) &= e^{2x}, \\
v_1(x) &= \frac{1}{4}e^{2x} - \frac{1}{4} - \frac{1}{2}\int_0^x v_0(t)dt = 0.
\end{aligned} \tag{105}$$

The exact solutions are therefore given by

$$u(x) = \pm e^x. \tag{106}$$

Example 3. Convert the Volterra integral equation of the first kind to the second kind and solve the resulting equation

$$\frac{1}{3}\sin x + \frac{1}{3}\sin(2x) = \int_0^x \cos(x - t)u^2(t)dt. \tag{107}$$

This equation is equivalent to

$$\frac{1}{3}\sin x + \frac{1}{3}\sin(2x) = \int_0^x \cos(x - t)v(t)dt, \tag{108}$$

upon setting $u^2(x) = v(x)$. Differentiating both sides of (108) and using Leibniz rule we obtain

$$v(x) = \frac{1}{3}\cos x + \frac{2}{3}\cos(2x) + \int_0^x \sin(x - t)v(t)dt. \tag{109}$$

For this problem, we select the Adomian decomposition method. Consequently, we use the recursion relation

$$
\begin{aligned}
v_0(x) &= \tfrac{1}{3}\cos x + \tfrac{2}{3}\cos(2x), \\
v_1(x) &= \tfrac{1}{6}x\sin x + \tfrac{2}{9}\cos(x) - \tfrac{2}{9}\cos(2x), \\
v_2(x) &= \tfrac{11}{72}x\sin x - \tfrac{1}{24}x^2\cos(x) - \tfrac{2}{27}\cos x + \tfrac{2}{27}\cos(2x),
\end{aligned}
\tag{110}
$$

$$\vdots$$

Recall that $u(x) = u_0 + u_1 + u_2 + \cdots$. Using the Taylor series for the sum of the obtained components gives

$$
u(x) = 1 - x^2 + \frac{1}{3}x^4 - \frac{2}{45}x^6 + \cdots,
\tag{111}
$$

which gives

$$
v(x) = \cos^2 x.
\tag{112}
$$

Consequently, the exact solution of the integral equation is given by

$$
u(x) = \pm\cos x,
\tag{113}
$$

obtained by noting that $v(x) = u^2(x)$.

Exercises 8.3.2

Convert the following nonlinear Volterra integral equations of the first kind to a second kind then solve the resulting equation:

1. $\dfrac{1}{7}x^7 + \dfrac{1}{56}x^8 = \displaystyle\int_0^x (x - t + 1)u^2(t)\,dt.$

2. $\dfrac{1}{9}x^9 + \dfrac{1}{90}x^{10} = \displaystyle\int_0^x (x - t + 1)u^2(t)\,dt.$

3. $\dfrac{1}{13}x^{13} + \dfrac{1}{182}x^{14} = \displaystyle\int_0^x (x - t + 1)u^3(t)\,dt.$

4. $\dfrac{3}{4}e^{2x} - \dfrac{1}{2}x - \dfrac{3}{4} = \displaystyle\int_0^x (x - t + 1)u^2(t)\,dt.$

5. $\dfrac{3}{4}x^2 e^{2x} + (\dfrac{5}{8} - x)e^{2x} - \dfrac{1}{4}x - \dfrac{5}{8} = \displaystyle\int_0^x (x - t + 1)u^2(t)\,dt.$

6. $\dfrac{1}{4}e^{-2x} + \dfrac{1}{2}x + \dfrac{1}{4} = \displaystyle\int_0^x (x - t + 1)u^2(t)\,dt.$

8.4   Nonlinear Weakly-Singular Volterra Integral Equations

In Chapter 6, we studied the linear weakly-singular Volterra integral equations of the second kind where the modified decomposition method was applied effectively. In this section we will study the nonlinear weakly-singular Volterra integral equations of the second kind.

The nonlinear weakly-singular Volterra integral equation of the second kind reads

$$u(x) = f(x) + \int_0^x \frac{1}{\sqrt{x-t}} F(u(t))dt, \ x \in [0,T], \tag{114}$$

and can be generalized to

$$u(x) = f(x) + \int_0^x \frac{1}{[g(x) - g(t)]^\alpha} F(u(t))dt, 0 < \alpha < 1, \ x \in [0,T], \tag{115}$$

where $F(u(t))$ is a nonlinear function of $u(t)$ such as $u^2(x)$ and $u^3(x)$, and the data function $f(x)$ is a given real-valued function. The unknown function $u(x)$ appears inside and outside the integral signs, a characteristic feature of a second-kind integral equation.

The nonlinear weakly-singular Volterra integral equations (114) and (115) arise in many mathematical physics and scientific applications such as stereology, heat conduction, crystal growth and the radiation of heat from a semi-infinite solid.

Numerical and analytical algorithms have been used for solving the nonlinear weakly-singular Volterra integral equations (114)–(115), where the main goal was the determination of numerical approximations of the solutions, or closed form solutions if exist. Examples of the methods that were used so far are the Galerkin collocation method, quadrature methods, spline methods, the homotopy analysis method, the Taylor series method, and the variational iteration method. Other useful methods such as the fast spectral method, the spectral method, the random point approximation method were also used.

In this work, we will use the Adomian decomposition method, or mostly the related techniques which are the modified Adomian decomposition method and the noise terms phenomenon to determine exact solutions of such nonlinear integral equations. In the sequel, we will briefly review the necessary steps of the two proposed schemes that will be used.

8.4.1 The Modified Decomposition Method

The Adomian method decomposes the solution $u(x)$ by the decomposition series

$$u(x) = \sum_{n=0}^{\infty} u_n(x), \tag{116}$$

and the nonlinear term $F(u(x))$ by the infinite series of Adomian polynomials

$$F(u(x)) = \sum_{n=0}^{\infty} A_n(x), \; A_n = \frac{1}{n!} \frac{d^n}{d\lambda^n} \left[F\left(\sum_{i=0}^{n} \lambda^i u_i \right) \right]_{\lambda=0} , n = 0, 1, 2, ...,$$

(117)

where the A_n are the Adomian polynomials, into both sides of the given equation. Consequently, the solution components $u_i(x), i \geq 0$, can be determined by using the standard recurrence relation

$$u_0(x) = f(x)$$

$$u_{n+1}(x) = \int_{a(x)}^{b(x)} K(x,t) A_n(t)\, dt, n \geq 1.$$

(118)

Having determined the solution components $u_0(x), u_1(x), u_2(x), \ldots$, the solution $u(x)$ follows immediately.

However, he modified decomposition method decomposes the data function $f(x)$ into two components $f_0(x)$ and $f_1(x)$, where only the first part is assigned to the zeroth solution component, and the latter is added to the first solution component $u_1(x)$ in addition to the other terms assigned by using the standard ADM. This in turn gives the modified recurrence relation by

$$u_0(x) = f_0(x),$$

$$u_1(x) = f_1(x) + \int_{a(x)}^{b(x)} K(x,t) A_0(t)\, dt,$$

$$u_{n+1}(x) = \int_{a(x)}^{b(x)} K(x,t) A_n(t)\, dt, n \geq 1.$$

(119)

Notice that the at least one of the limits of integration, either $a(x)$ or $b(x)$, must be a function of the independent variable in the case of Volterra integral equations. The use of the modified decomposition method not only minimizes the computations, but avoids the use of the higher order Adomian polynomials for such cases. It is worth noting that a proper selection of $f_0(x)$ and $f_1(x)$ is essential for a successful use of the modified decomposition method.

Moreover, the noise terms, if appearing especially within both of the components $u_0(x)$ and $u_1(x)$, will provide the exact solution by using only the first two iterations. By canceling the noise terms for $u_0(x)$, the remaining non-canceled terms of $u_0(x)$ may give the exact solution, and this can be verified through substitution into the original equation.

In what follows, we will examine several illustrative examples, where each one will be investigated by using the modified decomposition method, then by using the ADM combined with the noise terms phenomenon. In both cases, we only need the zeroth-order Adomian polynomial $A_0(x) = F(u_0(x))$. As will be shown in the sequel, there is no need to use the higher-order Adomian polynomials in such cases, because we shall see that the computations will mostly depend on the determination of $u_0(x)$ and $u_1(x)$.

In this section, we will study three nonlinear weakly-singular Volterra integral equations. We begin by applying the modified decomposition method and next by using the ADM combined with the noise terms phenomenon.

Example 1. Consider the nonlinear weakly-singular Volterra integral equation

$$u(x) = x^2 - \frac{3}{16}\pi x^4 + \int_0^x \frac{u^2(t)}{\sqrt{x^2 - t^2}}\, dt, 0 \le x \le 1. \qquad (120)$$

The modified decomposition method decomposes $f(x) = x^2 - \frac{3}{16}\pi x^4$ into two parts defined as

$$\begin{aligned} f_0(x) &= x^2, \\ f_1(x) &= -\frac{3}{16}\pi x^4. \end{aligned} \qquad (121)$$

Consequently, we can set the modified recurrence relation as

$$\begin{aligned} u_0(x) &= x^2, \\ u_1(x) &= -\frac{3}{16}\pi x^6 - \int_0^x \frac{A_0(t)}{\sqrt{x^2 - t^2}}\, dt = 0, \end{aligned} \qquad (122)$$

where the zeroth-order Adomian polynomial $A_0(x) = u_0^2(x)$. The subsequent solution components $u_j(x) = 0$ for $j \ge 2$. This gives the exact solution by

$$u(x) = x^2. \qquad (123)$$

However, using the ADM combined gives the standard recurrence relation

$$\begin{aligned} u_0(x) &= x^2 - \frac{3}{16}\pi x^4, \\ u_1(x) &= -\int_0^x \frac{A_0(t)}{\sqrt{x^2 - t^2}}\, dt, \\ &= \frac{3}{16}\pi x^4 + \text{other terms.} \end{aligned} \qquad (124)$$

The noise terms $\mp\frac{3}{16}\pi x^4$ appear within both $u_0(x)$ and $u_1(x)$. By canceling the noise term from $u_0(x)$ and verifying that the remaining non-canceled term in $u_0(x)$ identically satisfies the original equation (120), the exact solution is therefore given as

$$u(x) = x^2. \tag{125}$$

Example 2. Consider the nonlinear weakly-singular Volterra integral equation

$$u(x) = \sqrt[3]{1+x^2} - \frac{\pi}{2}(1+x^2) + \int_0^x \frac{u^3(t)}{\sqrt{x^2 - t^2}} \, dt, 0 \le x \le 1. \tag{126}$$

The modified decomposition method decomposes $f(x)$ into two parts defined as

$$\begin{aligned} f_0(x) &= \sqrt[3]{1+x^2}, \\ f_1(x) &= -\frac{\pi}{2}(1+x^2). \end{aligned} \tag{127}$$

Consequently, we can set the modified recurrence relation as

$$\begin{aligned} u_0(x) &= \sqrt[3]{1+x^2}, \\ u_1(x) &= -\frac{\pi}{2}(1+x^2) + \int_0^x \frac{A_0(t)}{\sqrt{x^2 - t^2}} \, dt \\ &= 0, \end{aligned} \tag{128}$$

where the zeroth-order Adomian polynomial $A_0(x) = u_0^3(x)$.

The exact solution,

$$u(x) = \sqrt[3]{1+x^2}, \tag{129}$$

follows immediately.

However, ADM admits the use of the standard recurrence relation

$$\begin{aligned} u_0(x) &= \sqrt[3]{1+x^2} - \frac{\pi}{2}(1+x^2), \\ u_1(x) &= \int_0^x \frac{A_0(t)}{\sqrt{x^2 - t^2}} \, dt \\ &= \frac{\pi}{2}(1+x^2) + \text{other terms}. \end{aligned} \tag{130}$$

The noise terms $\pm\frac{\pi}{2}(1+x^2)$ appear within both $u_0(x)$ and $u_1(x)$. Proceeding as before gives the same exact solution obtained earlier.

Example 3. Consider the nonlinear weakly-singular Volterra integral equation

$$u(x) = \sin^{\frac{1}{3}} x - \frac{3}{2}(\cos x - 1)^{\frac{2}{3}} + \int_0^x \frac{u^3(t)}{(\cos x - \cos t)^{\frac{1}{3}}} \, dt, 0 \le x \le 1. \tag{131}$$

The modified decomposition method decomposes $f(x)$ into two parts defined by

$$\begin{aligned} f_0(x) &= \sin^{\frac{1}{3}} x, \\ f_1(x) &= -\frac{3}{2}(\cos x - 1)^{\frac{2}{3}}. \end{aligned} \tag{132}$$

Consequently, we can set the modified recurrence relation as

$$\begin{aligned} u_0(x) &= \sin^{\frac{1}{3}} x, \\ u_1(x) &= -\frac{3}{2}(\cos x - 1)^{\frac{2}{3}} + \int_0^x \frac{A_0(t)}{(\sin x - \sin t)^{\frac{1}{3}}} \, dt = 0. \end{aligned} \tag{133}$$

The exact solution is thus given as

$$u(x) = \sin^{\frac{1}{3}} x. \tag{134}$$

Exercises 8.4

Solve the nonlinear weakly-singular Volterra integral equations

1. $u(x) = x^3 - \dfrac{729}{3080} x^{\frac{28}{3}} + \displaystyle\int_0^x \dfrac{u^3(t)}{(x^2 - t^2)^{\frac{1}{3}}} \, dt.$

2. $u(x) = x^4 - \dfrac{231}{2048} \pi x^{12} + \displaystyle\int_0^x \dfrac{u^3(t)}{\sqrt{x^2 - t^2}} \, dt.$

3. $u(x) = \sqrt[5]{1 + x^4} - \dfrac{\pi}{2}(1 + \dfrac{3}{8} x^4) + \displaystyle\int_0^x \dfrac{u^5(t)}{\sqrt{x^2 - t^2}} \, dt.$

4. $u(x) = \sqrt[4]{1 + x + x^3} - (\dfrac{\pi}{2} + x + \dfrac{2}{3} x^3) + \displaystyle\int_0^x \dfrac{u^4(t)}{\sqrt{x^2 - t^2}} \, dt.$

5. $u(x) = \sqrt[5]{\cos x} - 2\sqrt{\sin x} + \displaystyle\int_0^x \dfrac{u^5(t)}{\sqrt{\sin x - \sin t}} \, dt.$

6. $u(x) = \sqrt[4]{\sin x} + 2\sqrt{\cos x - 1} + \displaystyle\int_0^x \dfrac{u^4(t)}{\sqrt{\cos x - \cos t}} \, dt.$

Chapter 9

Applications of Integral Equations

9.1 Introduction

Several scientific and engineering applications are usually described by integral equations or integro differential equations, standard or singular. In Chapter 1, we have seen a large class of initial and boundary value problems that can be converted to Volterra or Fredholm integral equations. Integral equations arise in the potential theory more than any other field. Integral equations arise also in diffraction problems, conformal mapping, water waves, scattering in quantum mechanics, and Volterra's population growth model. The electrostatic, electro magnetic scattering problems and propagation of acoustical and elastical waves are scientific fields where integral equations appear.

We have presented in this text a variety of traditional methods and some newly developed methods to handle integral and integro-differential equations, Fredholm or Volterra type. Our concern in this text was the determination of the exact solutions in an easy computable fashion. Moreover, our aim was to present these selected methods to facilitate the computational work, and we avoided the abstract theorems that can be found in many other texts.

It is the goal of this chapter to select some applications that include integral or integro-differential equations. We will employ the methods introduced in this text. For numerical purposes we will use the Padé approximants that represent a function by a ratio of two polynomials.

9.2   Volterra Integral Form of the Lane-Emden Equation

The Lane-Emden equation appears mostly in astrophysics, such as the theory of stellar structure, and the thermal behavior of a spherical cloud of gas. The Lane-Emden equation comes in two kinds, namely the Lane-Emden equation of the first kind, or of index m, that reads

$$y'' + \frac{k}{x}y' + y^m = 0, y(0) = 1, y'(0) = 0, k > 1, \tag{1}$$

and the Lane-Emden equation of the second kind in the form

$$y'' + \frac{k}{x}y' + e^y = 0, y(0) = y'(0) = 0, k > 1. \tag{2}$$

The Lane-Emden equation of the first kind appears in astrophysics and used for computing the structure of interiors of polytropic stars, where exact solutions exist for $m = 0, 1, 5$. However, the Lane-Emden equation of the second kind models the non-dimensional density distribution $y(x)$ in an isothermal gas sphere. The singular behavior of these equations that occurs at $x = 0$ is the main difficulty of these two equations. Several methods were used in the literature to overcome the difficulty of the singular behavior.

A new study in [65] introduced a useful work to overcome the singularity behavior, where the Lane-Emden equation was converted to an equivalent Volterra integral form or to an equivalent Volterra integro-differential form. In what follows we summarize the formulation of new forms.

The generalized Lane-Emden equation of shape factor of $k > 1$ reads

$$y'' + \frac{k}{x}y' + f(y) = 0, y(0) = \alpha, y'(0) = 0, k > 1, \tag{3}$$

where $f(y)$ can take y^m or $e^{y(x)}$ as given earlier. To convert (3) to an integral form, it was given that

$$y(x) = \alpha - \frac{1}{k-1} \int_0^x t \left(1 - \frac{t^{k-1}}{x^{k-1}}\right) f(y(t)) \, dt. \tag{4}$$

Differentiating (4) twice, using the Leibniz rule, gives

$$
\begin{aligned}
y'(x) &= -\int_0^x \left(\frac{t^k}{x^k}\right) f(y(t)) \, dt, \\
y''(x) &= -f(y(x)) + \int_0^x k \left(\frac{t^k}{x^{k+1}}\right) f(y(t)) \, dt,
\end{aligned}
\tag{5}
$$

that can be proved by multiplying y' by $\frac{k}{x}$ and adding the result to $y''(x)$. This shows that the Lane-Emden equation (3) is equivalent to the Volterra integral form (4) or to the two Volterra integro-differential forms in (5) of first-order and second-order respectively.

It is to be noted that equations (4) and (5) work for any function $f(y)$, such as y^m and e^y as in the first kind and the second kind of the Lane-Emden equation. Moreover, $f(y)$ can be any linear or nonlinear function of y. A significant feature of equations (4) and (5) is the overcome of the singular behavior at $x = 0$. This was tested in [65] and proved to be reliable and efficient.

Moreover, for $k = 1$ the integral form is

$$y(x) = \alpha + \int_0^x t \ln(\frac{t}{x}) f(y(t)) \, dt, \tag{6}$$

which can be obtained by setting $k \to 1$ in Eq. (4).

Based on the last results we set the Volterra integral forms for the Lane–Emden equations as

$$y(x) = \begin{cases} \alpha + \displaystyle\int_0^x t \ln(\frac{t}{x}) f(y(t)) \, dt & \text{for } k = 1, \\[3mm] \alpha - \dfrac{1}{k-1} \displaystyle\int_0^x t \left(1 - \dfrac{t^{k-1}}{x^{k-1}}\right) f(y(t)) \, dt & \text{for } k > 1. \end{cases} \tag{7}$$

In what follows, we will examine the Lane-Emden equation of the first and the second kind, where the standard shape factor $k = 2$ will be used. For comparison reasons, we will use the Adomian decomposition method for the Volterra integral form (4). We then will apply the variational iteration method for the Volterra integro-differential form (5), because the variational iteration method is applicable to the integro-differential forms.

9.2.1 Lane-Emden Equation of the First Kind

The Lane-Emden equation of the first kind, or of index m, is

$$y'' + \frac{2}{x}y' + y^m = 0, \; y(0) = 1, y'(0) = 0, \tag{8}$$

which is a basic equation in the theory of stellar structure. This equation describes the temperature variation of a spherical gas cloud under the mutual attraction of its molecules and subject to the laws of thermodynamics. We will first start by using the Adomian decomposition method.

(i) By using the Adomian decomposition method:
Because the Lane-Emden equation involves the nonlinear term y^m, it is

normal to set the Adomian polynomials for the nonlinear term y^m by

$$
\begin{aligned}
A_0 &= y_0^m, \\
A_1 &= m y_0^{m-1} y_1, \\
A_2 &= m y_0^{m-1} y_2 + \frac{1}{2!} m(m-1) y_0^{m-2} y_1^2, \\
A_3 &= m y_0^{m-1} y_3 + m(m-1) y_0^{m-2} y_1 y_2 + \frac{1}{3!} m(m-1)(m-2) y_0^{m-3} y_1^3, \\
&\vdots
\end{aligned}
\tag{9}
$$

and so on for higher order polynomials.

Using (4) for $k = 2$, the recursive relation becomes

$$
\begin{aligned}
y_0(x) &= 1, \\
y_{n+1}(x) &= -\int_0^x t\left(1 - \frac{t}{x}\right) A_n \, dt, \ n \geq 0.
\end{aligned}
\tag{10}
$$

Using this relation, together with the Adomian polynomials (9) leads to the first few components

$$
\begin{aligned}
y_0(x) &= 1, \\
y_1(x) &= -\frac{1}{3!} x^2, \\
y_2(x) &= \frac{m}{4!} x^4, \\
y_3(x) &= -\frac{m(8m-5)}{3 \cdot 7!} x^6, \\
y_4(x) &= \frac{m(70 - 183m + 122m^2)}{9 \cdot 9!} x^8, \\
y_5(x) &= -\frac{m(3150 - 1080m + 12642m^2 - 5032m^3)}{45 \cdot 11!} x^{10}, \\
&\cdots
\end{aligned}
$$

Consequently, the generalized solution of the standard Lane-Emden equation takes the form

$$
\begin{aligned}
y(x) &= 1 - \frac{1}{3!} x^2 + \frac{m}{4!} x^4 - \frac{m(8m-5)}{3 \cdot 7!} x^6 + \frac{m(70 - 183m + 122m^2)}{9 \cdot 9!} x^8 \\
&\quad - \frac{m(3150 - 1080m + 12642m^2 - 5032m^3)}{45 \cdot 11!} x^{10} + \cdots.
\end{aligned}
\tag{11}
$$

The following exact solutions

$$y(x) = 1 - \frac{1}{3!}x^2,$$
$$y(x) = \frac{\sin x}{x},$$
$$y(x) = \left(1 + \frac{x^2}{3}\right)^{-\frac{1}{2}},$$

(12)

are obtained for $m = 0, 1$ and 5, respectively.

Fig. 1 below shows the Padé approximants $[4/4]$ of $y(x)$ in (11) for $m = 2, 3, 4, 5$. It is well-known that Padé approximants has the advantage of manipulating a polynomial approximation into a rational function to gain more information about the solution $y(x)$.

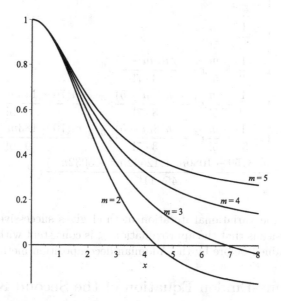

Fig. 1. The Padé approximants $[4/4]$ of $y(x))$ for $m = 2, 3, 4, 5$.

(ii) By using the variational iteration method:

Recall that the variational iteration method does not require the use of the Adomian polynomials. The variational iteration method can be used through a correction functional that requires the use of Lagrange multiplier λ. We first start with the generalized Lane-Emden equation of the first kind

$$y'' + \frac{k}{x}y' + y^m = 0, k \geq 1, y(0) = 1, y'(0) = 0.$$

(13)

Using (5) gives the integro-differential form of the Lane-Emden equation by

$$y'(x) = -\int_0^x \left(\frac{t^2}{x^2}\right) y^m(t)\, dt, y(0) = 1. \tag{14}$$

Consequently, the correction functional reads

$$y_{n+1}(x) = y_n(x) - \int_0^x \left(y_n'(t) + \int_0^t \frac{r^2}{t^2} y^m(r)\, dr\right) dt, y(0) = 1. \tag{15}$$

By selecting the zeroth selection $y_0(x) = 1$ we obtain the first few solution components

$$
\begin{aligned}
y_0(x) &= 1, \\
y_1(x) &= 1 - \frac{1}{3!}x^2, \\
y_2(x) &= 1 - \frac{1}{3!} + \frac{m}{5!}x^4, \\
y_3(x) &= 1 - \frac{1}{3!} + \frac{m}{5!}x^4 - \frac{m(8m-5)}{3\cdot 7!}x^6, \\
y_4(x) &= 1 - \frac{1}{3!} + \frac{m}{5!}x^4 - \frac{m(8m-5)}{3\cdot 7!}x^6 + \frac{m(70-183m+122m^2}{9\cdot 9!}x^8 \\
y_5(x) &= 1 - \frac{1}{3!} + \frac{m}{5!}x^4 - \frac{m(8m-5)}{3\cdot 7!}x^6 + \frac{m(70-183m+122m^2}{9\cdot 9!}x^8 \\
&\quad - \frac{m(3150-1080m+12642m^2-5032m^3)}{45\cdot 11!}x^{10} \\
&\quad \cdots.
\end{aligned}
$$

Recall that the variational iteration method gives successive approximations. This means that the approximation y_5 is consistent with the approximation obtained before by the Adomian decomposition method.

9.2.2 Lane-Emden Equation of the Second Kind

In this section, we will study the Lane-Emden equation of the second kind

$$y'' + \frac{2}{x}y' + e^y = 0, y(0) = 0, y'(0) = 0 \tag{16}$$

by using the Adomian decomposition method and then by the variational iteration method. This equation, models the distribution of mass in an isothermal gas sphere.

(i) By using the Adomian decomposition method:
Because the Lane-Emden equation involves the nonlinear term $e^{y(x)}$, it is

normal to set the Adomian polynomials for the nonlinear term $e^{y(x)}$ by

$$
\begin{aligned}
A_0 &= e^{y_0}, \\
A_1 &= y_1 e^{y_0}, \\
A_2 &= \left(\frac{1}{2!} y_1^2 + y_2 \right) e^{y_0}, \\
A_3 &= \left(\frac{1}{3!} y_1^3 + y_1 y_2 + y_3 \right) e^{y_0}, \\
A_4 &= \left(\frac{1}{4!} y_1^4 + \frac{1}{2!} (y_1^2 y_2^2) + y_1 y_3 + y_4 \right) e^{y_0}, \\
&\cdots.
\end{aligned}
\tag{17}
$$

To solve the Lane-Emden equation (16) by using the Adomian decomposition method, we substitute $k = 2$, $\alpha = 0$ and $f(y) = e^y$ in (4), and using the recursive relation

$$
\begin{aligned}
y_0(x) &= 0, \\
y_{n+1}(x) &= -\int_0^x t \left(1 - \frac{t}{x} \right) A_n \, dt, \; n \geq 0,
\end{aligned}
\tag{18}
$$

as presented before. Using (18) gives the following components

$$
\begin{aligned}
y_0(x) &= 0, \\
y_1(x) &= -\frac{1}{3!} x^2, \\
y_2(x) &= \frac{1}{5!} x^4, \\
y_3(x) &= -\frac{8}{3 \times 7!} x^6, \\
y_4(x) &= \frac{122}{9 \times 9!} x^8, \\
y_5(x) &= -\frac{5032}{45 \times 11!} x^{10}, \\
&\cdots.
\end{aligned}
\tag{19}
$$

The series solution is therefore given by

$$
y(x) = -\frac{1}{3!} x^2 + \frac{1}{5!} x^4 - \frac{8}{3 \times 7!} x^6 + \frac{122}{9 \times 9!} x^8 - \frac{5032}{45 \times 11!} x^{10} + \cdots. \tag{20}
$$

Fig. 2 below shows the Padé approximants $-[4/4]$ and $-[5/5]$ of $y(x)$ in (20).

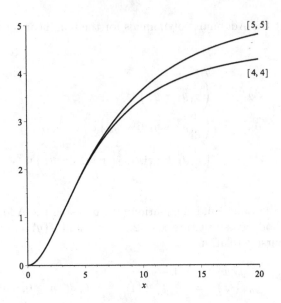

Fig. 2.   The Padé approximants $-[4/4]$ and $-[5/5]$ of $y(x)$ in (20).

(ii) By using the variational iteration method:

To solve the Lane-Emden equation of the second kind, we set the correction functional

$$y_{n+1}(x) = y_n(x) - \int_0^x \left(y_n'(t) + \int_0^t \frac{r^2}{t^2} e^{y_n(r)} \right) dt, \quad y(0) = 0. \qquad (21)$$

Proceeding as in the previous case, we obtain the following components

$$
\begin{aligned}
y_0(x) &= 0, \\
y_1(x) &= -\frac{1}{3!}x^2, \\
y_2(x) &= -\frac{1}{3!}x^2 + \frac{1}{5!}x^4 + \cdots, \\
y_3(x) &= -\frac{1}{3!}x^2 + \frac{1}{5!}x^4 - \frac{8}{3 \times 7!}x^6 + \cdots, \\
y_4(x) &= -\frac{1}{3!}x^2 + \frac{1}{5!}x^4 - \frac{8}{3 \times 7!}x^6 + \frac{122}{9 \times 9!}x^8, \\
y_5(x) &= -\frac{1}{3!}x^2 + \frac{1}{5!}x^4 - \frac{8}{3 \times 7!}x^6 + \frac{122}{9 \times 9!}x^8 - \frac{5032}{45 \times 11!}x^{10},
\end{aligned}
\qquad (22)
$$

$$\vdots$$

The series solution is therefore given by

$$y(x) = -\frac{1}{3!}x^2 + \frac{1}{5!}x^4 - \frac{8}{3 \times 7!}x^6 + \frac{122}{9 \times 9!}x^8 - \frac{5032}{45 \times 11!}x^{10} + \cdots, \quad (23)$$

which is consistent with the result obtained by using the Adomian decomposition method.

In closing this section, it is worth noting that the Adomian decomposition method requires the use of Adomian polynomials for nonlinear terms that needs more computational work. However, using the variational iteration method requires the use of Lagrange multiplier λ, where we used $\lambda = -1$ for first order integro-differential equations.

9.3 The Schlömilch's Integral Equation

In this section we will study the Schlömilch's integral equation that was used for the computation of ionospheric height that corresponds to a given plasma frequency from oblique propagation data. Although the Schlömilch's integral equation is linear, but we will study the linear and the nonlinear cases.

9.3.1 The Linear Schlömilch's Integral Equation

The Schlömilch's integral equation reads

$$f(x) = \frac{2}{\pi} \int_0^{\frac{\pi}{2}} u(x \sin t) \, dt, \quad (24)$$

where $f(x)$ is a continuous differential coefficient for $-\pi \leq x \leq \pi$. It has been proved that this equation gives one solution given by

$$u(x) = f(0) + x \int_0^{\frac{\pi}{2}} f'(x \sin t) \, dt, \quad (25)$$

where f' is the derivative of f with respect to the complicated argument $\xi = x \sin t$. The Schlömilch's integral equation and its unique solution have been used to determine the electron density profile from the ionospheric ionograms for the case of the quasi-transverse (QT) approximations [41]. However, our study on the Schlömilch's integral equation will be focused on the mathematical side only.

It is obvious that the Schlömilch's integral equation is a special case of Fredholm integral equation of the first kind. The method of regularization was used effectively in Chapter 2 to handle the Fredholm integral equation of the first kind. In view of this, we will employ the method of regularization to handle the Schlömilch's integral equation (24).

9.3.2   The Method of Regularization

The method of regularization was introduced in Chapter 2, hence we skip details. The method of regularization converts the linear Schlömilch's integral equation of the first kind

$$f(x) = \frac{2}{\pi} \int_0^{\frac{\pi}{2}} u(x \sin t)\, dt, x \in D, \tag{26}$$

to the Schlömilch's integral equation of the second kind in the form

$$\epsilon u_\epsilon(x) = f(x) - \frac{2}{\pi} \int_0^{\frac{\pi}{2}} u_\epsilon(x \sin t)\, dt, \tag{27}$$

or equivalently

$$u_\epsilon(x) = \frac{1}{\epsilon} f(x) - \frac{1}{\epsilon} \left(\frac{2}{\pi} \int_0^{\frac{\pi}{2}} u_\epsilon(x \sin t)\, dt \right), \tag{28}$$

where ϵ is a small positive parameter. It is obvious that the solution u_ϵ of equation (28) converges to the solution $u(x)$ of (26) as $\epsilon \to 0$. Consequently, we can apply any method that we studied before for solving the Fredholm integral equations of the second kind. However, we will use the Adomian decomposition method, where we set the recurrence relation as

$$
\begin{aligned}
u_{\epsilon_0}(x) &= \frac{1}{\epsilon} f(x), \\
u_{\epsilon_{k+1}}(x) &= -\frac{1}{\epsilon} \left(\frac{2}{\pi} \int_0^{\frac{\pi}{2}} u_{\epsilon_k}(x \sin t)\, dt \right), k \geq 0.
\end{aligned}
\tag{29}
$$

In what follows we will present two illustrative examples. Our focus will be on transforming the first kind equation to a second kind equation by using the method of regularization, and hence we can use any appropriate method.

Example 1. Solve the Schlömilch's integral equation

$$1 + x = \frac{2}{\pi} \int_0^{\frac{\pi}{2}} u(x \sin t)\, dt, -\pi \leq x \leq \pi. \tag{30}$$

Using the method of regularization, Eq. (30) becomes

$$u_\epsilon(x) = \frac{1}{\epsilon}(1 + x) - \frac{1}{\epsilon} \left(\frac{2}{\pi} \int_0^{\frac{\pi}{2}} u_\epsilon(x \sin t)\, dt \right). \tag{31}$$

We select the Adomian method for solving this equation. The Adomian method admits the use of the recurrence relation

$$u_{\epsilon_0}(x) = \frac{1}{\epsilon}(1+x),$$

$$u_{\epsilon_{k+1}}(x) = -\frac{1}{\epsilon}\left(\frac{2}{\pi}\int_0^{\frac{\pi}{2}} u_{\epsilon_k}(x\sin t)\,dt\right), k \geq 0. \tag{32}$$

This in turn gives the first few components

$$u_{\epsilon_0}(x) = \frac{1}{\epsilon}(1+x),$$

$$u_{\epsilon_1}(x) = -\frac{\pi+2x}{\pi\epsilon^2},$$

$$u_{\epsilon_2}(x) = \frac{\pi^2+4x}{\pi^2\epsilon^3},$$

$$u_{\epsilon_3}(x) = -\frac{\pi^3+8x}{\pi^3\epsilon^4}, \tag{33}$$

$$u_{\epsilon_4}(x) = \frac{\pi^4+16x}{\pi^4\epsilon^5},$$

$$\vdots$$

This in turn gives

$$u_\epsilon(x) = \frac{1}{\epsilon}\left(1-\frac{1}{\epsilon}+\frac{1}{\epsilon^2}-\frac{1}{\epsilon^3}+\cdots\right) + \frac{x}{\epsilon}\left(1-\frac{2}{\pi\epsilon}+\frac{4}{\pi^2\epsilon^2}-\frac{8}{\pi^3\epsilon^3}+\cdots\right), \tag{34}$$

which gives

$$u_\epsilon(x) = \frac{1}{1+\epsilon} + \frac{\pi}{2+\pi\epsilon}x, \tag{35}$$

obtained upon finding the summation of each infinite geometric series. The exact solution is given by

$$u(x) = \lim_{\epsilon\to 0} u_\epsilon(x) = 1 + \frac{\pi}{2}x. \tag{36}$$

Example 2. Solve the Schlömilch's integral equation

$$3x^4 = \frac{2}{\pi}\int_0^{\frac{\pi}{2}} u(x\sin t)\,dt, -\pi \leq x \leq \pi. \tag{37}$$

Using the method of regularization, Eq. (37) becomes

$$u_\epsilon(x) = \frac{3}{\epsilon}x^4 - \frac{1}{\epsilon}\left(\frac{2}{\pi}\int_0^{\frac{\pi}{2}} u_\epsilon(x\sin t)\,dt\right). \tag{38}$$

We select the Adomian method for solving this equation. The Adomian method admits the use of the recurrence relation

$$u_{\epsilon_0}(x) = \frac{3}{\epsilon}x^4,$$

$$u_{\epsilon_{k+1}}(x) = -\frac{1}{\epsilon}\left(\frac{2}{\pi}\int_0^{\frac{\pi}{2}} u_{\epsilon_k}(x\sin t)\,dt\right), k \geq 0. \tag{39}$$

This in turn gives the first few components

$$u_{\epsilon_0}(x) = \frac{3}{\epsilon}x^4,$$

$$u_{\epsilon_1}(x) = -\frac{9}{8\epsilon^2}x^4,$$

$$u_{\epsilon_2}(x) = \frac{27}{64\epsilon^3}x^4, \tag{40}$$

$$u_{\epsilon_3}(x) = -\frac{81}{512\epsilon^4}x^4.$$

This in turn gives

$$u_\epsilon(x) = \frac{3}{\epsilon}x^4\left(1 - \frac{3}{8\epsilon} + \frac{9}{64\epsilon^2} - \frac{27}{512\epsilon^3} + \cdots\right), \tag{41}$$

which gives

$$u_\epsilon(x) = \frac{24}{3+8\epsilon}x^4, \tag{42}$$

obtained upon finding the summation of the infinite geometric series. The exact solution is given by

$$u(x) = \lim_{\epsilon\to 0} u_\epsilon(x) = 8x^4. \tag{43}$$

9.3.3   The Nonlinear Schlömilch's Integral Equation

The Schlömilch's integral equation may come in a nonlinear form as

$$f(x) = \frac{2}{\pi}\int_0^{\frac{\pi}{2}} F(u(x\sin t))\,dt, \tag{44}$$

where $F(u(x\sin t))$ is a nonlinear function of $u(x\sin t)$, such as $u^2(x\sin t)$ and $u^3(x\sin t)$, $f(x)$ is a continuous differential coefficient for $-\pi \leq x \leq \pi$.

It is obvious that the nonlinear Schlömilch's integral equation is a special case of the nonlinear Fredholm integral equation of the first kind. It is thus normal to follow the analysis used before in Chapter 7 for handling nonlinear Fredholm integral equation of the first kind.

We first transform (44) to a linear form of the first kind given by

$$f(x) = \frac{2}{\pi} \int_0^{\frac{\pi}{2}} v(x \sin t) \, dt, x \in D \tag{45}$$

by using the transformation

$$v(x \sin t) = F(u(x \sin t)). \tag{46}$$

Assuming that $F(u(x \sin t))$ is invertible leads to

$$u(x \sin t) = F^{-1}(v(x)). \tag{47}$$

The method of regularization converts the linear Schlömilch's integral equation of the first kind (45) to the Schlömilch's integral equation of the second kind in the form

$$v_\epsilon(x) = \frac{1}{\epsilon} f(x) - \frac{1}{\epsilon} \left(\frac{2}{\pi} \int_0^{\frac{\pi}{2}} v_\epsilon(x \sin t) \, dt \right), \tag{48}$$

where ϵ is a small positive parameter. Applying the Adomian decomposition method gives the recurrence relation

$$
\begin{aligned}
v_{\epsilon_0}(x) &= \frac{1}{\epsilon} f(x), \\
v_{\epsilon_{k+1}}(x) &= -\frac{1}{\epsilon} \left(\frac{2}{\pi} \int_0^{\frac{\pi}{2}} v_{\epsilon_k}(x \sin t) \, dt \right), k \geq 0.
\end{aligned}
\tag{49}
$$

The scheme that we presented will be illustrated by examining the following examples.

Example 3. Solve the Schlömilch's integral equation

$$\frac{3}{8} x^4 = \frac{2}{\pi} \int_0^{\frac{\pi}{2}} u^2(x \sin t) \, dt, -\pi \leq x \leq \pi. \tag{50}$$

Using the transformation $v = u^2$ carries out the last equation to

$$\frac{3}{8} x^4 = \frac{2}{\pi} \int_0^{\frac{\pi}{2}} v(x \sin t) \, dt, -\pi \leq x \leq \pi. \tag{51}$$

Using the method of regularization, Eq. (51) becomes

$$v_\epsilon(x) = \frac{3}{8\epsilon} x^4 - \frac{1}{\epsilon} \left(\frac{2}{\pi} \int_0^{\frac{\pi}{2}} v_\epsilon(x \sin t) \, dt \right). \tag{52}$$

The Adomian method gives the recurrence relation

$$
\begin{aligned}
v_{\epsilon_0}(x) &= \frac{3}{8\epsilon}x^4, \\
v_{\epsilon_{k+1}}(x) &= -\frac{1}{\epsilon}\left(\frac{2}{\pi}\int_0^{\frac{\pi}{2}} v_{\epsilon_k}(x\sin t)\,dt\right), k \geq 0.
\end{aligned}
\tag{53}
$$

This in turn gives the first few components

$$
\begin{aligned}
v_{\epsilon_0}(x) &= \frac{3}{8\epsilon}x^4, \\
v_{\epsilon_1}(x) &= -\frac{9}{64\epsilon^2}x^4, \\
v_{\epsilon_2}(x) &= \frac{27}{512\epsilon^3}x^4.
\end{aligned}
\tag{54}
$$

This in turn gives

$$
v_\epsilon(x) = \frac{3}{8\epsilon}x^4\left(1 - \frac{3}{8\epsilon} + \frac{9}{64\epsilon^2} + \cdots\right) + \frac{x}{\epsilon}\left(1 - \frac{2}{\pi\epsilon} + \frac{4}{\pi^2\epsilon^2} + \cdots\right),
\tag{55}
$$

which gives

$$
v_\epsilon(x) = \frac{3}{3 + 8\epsilon}x^4,
\tag{56}
$$

obtained upon finding the summation of each infinite geometric series. Recall that $v = u^2$, hence the exact solution is given by

$$
v(x) = x^4, u(x) = \pm x^2.
\tag{57}
$$

Example 4. Solve the Schlömilch's integral equation

$$
\frac{32}{3\pi}x^3 = \frac{2}{\pi}\int_0^{\frac{\pi}{2}} u^3(x\sin t)\,dt, -\pi \leq x \leq \pi.
\tag{58}
$$

Using the transformation $v = u^3$ carries out the last equation to

$$
\frac{32}{3\pi}x^3 = \frac{2}{\pi}\int_0^{\frac{\pi}{2}} v(x\sin t)\,dt, -\pi \leq x \leq \pi.
\tag{59}
$$

Using the method of regularization, Eq. (51) becomes

$$
v_\epsilon(x) = \frac{32}{3\pi}x^3 - \frac{1}{\epsilon}\left(\frac{2}{\pi}\int_0^{\frac{\pi}{2}} v_\epsilon(x\sin t)\,dt\right).
\tag{60}
$$

The Adomian method gives the recurrence relation

$$
\begin{aligned}
v_{\epsilon_0}(x) &= \frac{32}{3\pi\epsilon}x^3, \\
v_{\epsilon_{k+1}}(x) &= -\frac{1}{\epsilon}\left(\frac{2}{\pi}\int_0^{\frac{\pi}{2}} v_{\epsilon_k}(x\sin t)\,dt\right), k \geq 0.
\end{aligned}
\tag{61}
$$

This in turn gives the first few components

$$
\begin{aligned}
v_{\epsilon_0}(x) &= \frac{32}{3\pi\epsilon}x^3, \\
v_{\epsilon_1}(x) &= -\frac{128}{9\pi^2\epsilon^2}x^3, \\
v_{\epsilon_2}(x) &= \frac{512}{27\pi^3\epsilon^3}x^3, \\
v_{\epsilon_3}(x) &= -\frac{2048}{81\pi^4\epsilon^4}x^3.
\end{aligned}
\tag{62}
$$

Proceeding as before, and recall that $v = u^3$, hence the exact solution is given by

$$
v(x) = 8x^3, u(x) = 2x.
\tag{63}
$$

9.4 Bratu-Type Problems

It is well known that Bratu's boundary value problem in 1-dimensional planar coordinates is of the form

$$
\begin{aligned}
u'' + \lambda e^u &= 0, 0 < x < 1, \\
u(0) &= u(1) = 0.
\end{aligned}
\tag{64}
$$

The standard Bratu problem (64) was used to model a combustion problem in a numerical slab. The Bratu problem appears in many scientific applications such as the fuel ignition of the thermal combustion theory and in the Chandrasekhar model of the expansion of the universe. It stimulates a thermal reaction process in a rigid material where the process depends on a balance between chemically generated heat, radiative heat transfer and nanotechnology, and heat transfer by conduction.

The Bratu problem was subjected to a considerable amount of research work. Several numerical techniques, such as the finite difference method, finite element approximation, weighted residual method, Adomian-Laplace method, the variational iteration method, collocation method, Chebyshev wavelets method, and the shooting method have been implemented independently to handle the Bratu model numerically. In addition, Boyd [5] employed Chebyshev polynomial expansions and the Gegenbauer polynomials as base functions.

The exact solution to (64) reads [5]

$$u(x) = -2 \ln \left[\frac{\cosh((x - \frac{1}{2})\frac{\theta}{2})}{\cosh(\frac{\theta}{4})} \right], \tag{65}$$

where θ satisfies

$$\theta = \sqrt{2\lambda} \cosh \left(\frac{\theta}{4} \right). \tag{66}$$

The Bratu problem has zero, one or two solutions when $\lambda > \lambda_c, \lambda = \lambda_c$, and $\lambda < \lambda_c$ respectively, where the critical value λ_c satisfies the equation

$$\lambda_c = 8 \operatorname{csch}^2 \left(\frac{\theta_c}{4} \right). \tag{67}$$

It was evaluated in [5] that the critical value λ_c is given by

$$\lambda_c = 3.513830719. \tag{68}$$

In this section we will concern ourselves on studying three distinct Bratu-type problems, given by

$$\begin{aligned} u'' - \pi^2 e^u &= 0, \, 0 < x < 1, \\ u(0) &= u(1) = 0. \end{aligned} \tag{69}$$

$$\begin{aligned} u'' + \pi^2 e^{-u} &= 0, \, 0 < x < 1 \\ u(0) &= u(1) = 0 \end{aligned} \tag{70}$$

and

$$\begin{aligned} u'' - e^u &= 0, \, 0 < x < 1, \\ u(0) &= u(1) = 0. \end{aligned} \tag{71}$$

The last equation (71) is of great interest in magneto hydrodynamics [4]. The aforementioned Bratu-type problems were examined by many methods in the literature. In this section, we will convert these problems to equivalent Volterra integro-differential equations and study it by the methods used in this text.

9.4.1   First Bratu-Type Problem

We begin this analysis by studying the first Bratu-type problem

$$\begin{aligned} u'' - \pi^2 e^u &= 0, \, 0 < x < 1, \\ u(0) &= u(1) = 0. \end{aligned} \tag{72}$$

Integrating both sides of (72) converts this equation to an equivalent Volterra integro-differential equation given by

$$u' = a + \pi^2 \int_0^x e^{u(t)} \, dt \quad 0 < x < 1, \tag{73}$$

where $a = u'(0) \neq 0$ is not defined but will be determined by using the boundary condition $u(1) = 0$. Applying the variational iteration method for (73) gives the correction functional reads

$$u_{n+1}(x) = u_n(x) - \int_0^x \left(u_n'(t) - a - \pi^2 \int_0^t e^{u_n(r)} \, dr \right) dt. \tag{74}$$

By selecting the zeroth selection $u_0(x) = 0$ we obtain the first few solution components

$$
\begin{aligned}
u_0(x) &= 0, \\
u_1(x) &= ax + \frac{\pi^2}{2}x^2, \\
u_2(x) &= ax + \frac{\pi^2}{2}x^2 + \frac{a\pi^2}{6}x^3 + \left(\frac{\pi^4}{4!} + \frac{a^2\pi^2}{4!} \right)x^4 + \text{other terms}, \\
u_3(x) &= ax + \frac{\pi^2}{2}x^2 + \frac{a\pi^2}{6}x^3 + \left(\frac{\pi^4}{4!} + \frac{a^2\pi^2}{4!} \right)x^4 \\
&\quad + \left(\frac{a\pi^4}{30} + \frac{a^3\pi^2}{6!} \right)x^5 + \text{other terms},
\end{aligned}
\tag{75}
$$

where the Taylor series for the nonlinear term $e^{u(x)}$ was used in the computational work. To determine a, we need to use the Padé approximants. Padé approximants represent a function by the ratio of two polynomials. Padé approximant, symbolized by $[m/n]$, is a rational function defined by

$$[m/n] = \frac{a_0 + a_1 x + a_2 x^2 + \cdots + a_m x^m}{1 + b_1 x + b_2 x^2 + \cdots + b_n x^n}, \tag{76}$$

where the numerator and denominator have no common factors. If we selected $m = n$, then the approximants $[n/n]$ are called diagonal approximants. The coefficients of the polynomials in the numerator and in the denominator are determined by using the coefficients in the Taylor expansion of the function which we need to find its approximants. For example, considering $f(x) = e^x$, the Padé $[2/2]$ and $[3/3]$ approximants are given by

$$[2/2] = \frac{12 + 6x + x^2}{12 - 6x + x^2}, \quad [3/3] = \frac{120 + 60x + 12x^2 + x^3}{120 - 60x + 12x^2 - x^3}.$$

However, Padé approximants can be evaluated by using any symbolic computer software such as Maple or Mathematica.

To determine a, we use the boundary condition $u(1) = 0$ in the Padé approximants $[2/2], [3/3], [4/4], \ldots$, of (75) to find that $a = \pi$. Substituting $a = \pi$ in $u_3(x)$ gives the series solution

$$u(x) = \pi x + \frac{\pi^2}{2}x^2 + \frac{\pi^3}{6}x^3 + \frac{\pi^4}{12}x^4 + \frac{\pi^5}{24}x^5 + \cdots. \qquad (77)$$

This in turn gives the exact solution by

$$u(x) = -\ln\left(1 + \cos\left((\frac{1}{2} + x)\pi\right)\right). \qquad (78)$$

Fig. 3 below shows the solution $u(x)$ in (78) that blows up in the middle of the domain.

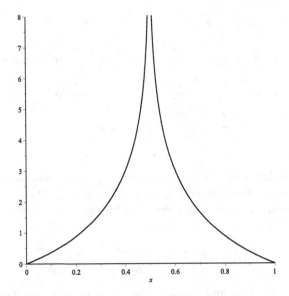

Fig. 3.   The solution $u(x)$ in (78).

9.4.2   Second Bratu-Type Problem

In this section we will the second Bratu-type problem

$$\begin{aligned} u'' + \pi^2 e^{-u} &= 0, \, 0 < x < 1, \\ u(0) &= u(1) = 0. \end{aligned} \qquad (79)$$

Integrating both sides of (79) converts this equation to an equivalent Volterra integro-differential equation given by

$$u' = a - \pi^2 \int_0^x e^{-u(t)} \, dt \, 0 < x < 1, \tag{80}$$

where $a = u'(0) \neq 0$ is not defined but will be determined by using the boundary condition $u(1) = 0$. Applying the variational iteration method for (80) gives the correction functional reads

$$u_{n+1}(x) = u_n(x) - \int_0^x \left(u'_n(t) - a + \pi^2 \int_0^t e^{-u_n(r)} \, dr \right) dt. \tag{81}$$

By selecting the zeroth selection $u_0(x) = 0$ we obtain the first few solution components

$$
\begin{aligned}
u_0(x) &= 0, \\
u_1(x) &= ax - \frac{\pi^2}{2}x^2, \\
u_2(x) &= ax - \frac{\pi^2}{2}x^2 + \frac{a\pi^2}{6}x^3 - (\frac{\pi^4}{4!} + \frac{a^2\pi^2}{4!})x^4 + \text{other terms}, \\
u_3(x) &= ax + \frac{\pi^2}{2}x^2 + \frac{a\pi^2}{6}x^3 + (\frac{\pi^4}{4!} + \frac{a^2\pi^2}{4!})x^4 \\
&\quad + (\frac{a\pi^4}{30} + \frac{a^3\pi^2}{6!})x^5 + \text{other terms},
\end{aligned}
\tag{82}
$$

where the Taylor series for the nonlinear term $e^{-u(x)}$ was used in the computational work.

To determine a, we use the boundary condition $u(1) = 0$ in the Padé approximants $[2/2], [3/3], [4/4], \ldots$, to find that $a = \pi$. Substituting $a = \pi$ in $u_3(x)$ gives the series solution

$$u(x) = \pi x - \frac{\pi^2}{2}x^2 + \frac{\pi^3}{6}x^3 - \frac{\pi^4}{12}x^4 + \frac{\pi^5}{24}x^5 + \cdots. \tag{83}$$

This in turn gives the exact solution by

$$u(x) = \ln\left(1 + \sin(\pi x)\right). \tag{84}$$

Fig. 4 below shows the exact solution $u(x)$ in (84) and the approximate solution (83). The graph shows that $u(x)$ is bounded, and the deviation between the exact solution and the approximation is caused by the use of finite number of terms of the series solution.

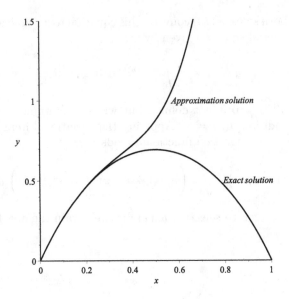

Fig. 4. The exact solution $u(x)$ in (84) and the approximate solution (83).

9.4.3   Third Bratu-Type Problem

We close this section on Bratu-type problems by examining a third Bratu-type problem in the form

$$
\begin{aligned}
u'' - e^u &= 0,\, 0 < x < 1, \\
u(0) &= u(1) = 0,
\end{aligned}
\tag{85}
$$

which is of great interest in magneto hydrodynamics [4]. Proceeding as before, we integrate both sides of (85) to convert it to an equivalent Volterra integro-differential equation as

$$
u' = a + \int_0^x e^{u(t)}\, dt,\, 0 < x < 1,
\tag{86}
$$

where $a = u'(0) \neq 0$ is not defined yet, but will be determined by using the boundary condition $u(1) = 0$. Applying the variational iteration method for (86) gives the correction functional by

$$
u_{n+1}(x) = u_n(x) - \int_0^x \left(u_n'(t) - a - \int_0^t e^{u_n(r)}\, dr \right) dt.
\tag{87}
$$

By selecting the zeroth selection $u_0(x) = 0$ we obtain the first few solution components

$$
\begin{aligned}
u_0(x) &= 0, \\
u_1(x) &= ax + \frac{1}{2}x^2, \\
u_2(x) &= ax + \frac{1}{2}x^2 + \frac{a}{6}x^3 + (\frac{1}{4!} + \frac{a^2}{4!})x^4 + \text{other terms}, \\
u_3(x) &= ax + \frac{1}{2}x^2 + \frac{a}{6}x^3 + (\frac{1}{4!} + \frac{a^2}{4!})x^4 \\
&\quad + (\frac{a}{30} + \frac{a^3}{6!})x^5 + \text{other terms}, \\
&\cdots ,
\end{aligned}
\tag{88}
$$

where the Taylor series for the nonlinear term $e^{u(x)}$ was used in the computational work. Using the boundary condition $u(1) = 0$ gives

$$
a = u'(0) = -0.463639988227675. \tag{89}
$$

Substituting this result into $u_3(x)$ gives the series approximation

$$
\begin{aligned}
u(x) &= -0.463639988227675x + 0.5x^2 - 0.07727333137x^3 \\
&+ 0.05062341828x^4 - 0.01628520791x^5 + 0.008903876534x^6 \\
&- 0.003646125737x^7 + 0.001868477052x^8 - 0.000851751709x^9 \\
&+ 0.0004283517518x^{10} - 0.0002057749479x^{11} \\
&+ 0.0001033495701x^{12} - 0.00005105422613x^{13} \\
&+ 0.00002576095109x^{14} + \cdots .
\end{aligned}
\tag{90}
$$

However, an exact solution is given by

$$
u(x) = -\ln 2 + \ln(\lambda(x)), \tag{91}
$$

where

$$
\lambda(x) = \left(c \sec \left(\frac{c(2x - 1)}{4} \right) \right)^2, \tag{92}
$$

and c is the root of

$$
\left(c \sec(\frac{c}{4}) \right)^2 = 2. \tag{93}
$$

The root c lies between 0 and $\frac{\pi}{2}$, namely $c = 1.336055695$, to ten figures.

9.5 Systems of Integral Equations

Systems of Fredholm and Volterra integral equations, of the first and the second kinds, appear in scientific applications. Many powerful methods were used to study these systems of equations. In this section, we aim to apply the methods presented in this text to study systems of integral equations.

9.5.1 Systems of Fredholm Integral Equations

In this section we will concern ourselves on solving system of Fredholm integral equations of the second kind with only two unknown functions $u(x)$ and $v(x)$. The analysis can be extended to more than two unknown functions. The standard system of Fredholm integral equations of the second kind is given by

$$
\begin{aligned}
u(x) &= f_1(x) + \int_a^b \left(K_1(x,t)u(t) + \tilde{K}_1(x,t)v(t) \right) dt, \\
v(x) &= f_2(x) + \int_a^b \left(K_2(x,t)u(t) + \tilde{K}_2(x,t)v(t) \right) dt,
\end{aligned}
\tag{94}
$$

where the unknown functions that will be determined are $u(x)$ and $v(x)$. Recall that for the second kind, the unknown functions appear inside and outside the integral sign. The kernels $K_i(x,t)$, and $\tilde{K}_i(x,t), i = 1,2$, and the function $f_1(x)$ and $f_2(x)$ are prescribed real-valued functions.

Recall that we applied the direct computation method and the Adomian decomposition method, together with its related modification, for handling Fredholm integral equations. The aforementioned methods are now well known, hence we will select two distinct examples that will be examined by using the two methods.

Example 1. Use the Adomian decomposition method and the noise terms phenomenon to solve the following system of Fredholm integral equations

$$
\begin{aligned}
u(x) &= x + \sin x - \frac{\pi^2}{4} - \frac{\pi}{2}x + \int_0^{\frac{\pi}{2}} \left((1+xt)u(t) + (1-xt)v(t) \right) dt, \\
v(x) &= x - \cos x - \frac{\pi^2}{4} + \frac{\pi}{2}x + \int_0^{\frac{\pi}{2}} \left((1-xt)u(t) + (1+xt)v(t) \right) dt.
\end{aligned}
\tag{95}
$$

The Adomian decomposition method decomposes $u(x)$ and $v(x)$ by an infinite series of components

$$
\begin{aligned}
u(x) &= \sum_{n=0}^{\infty} u_n(x), \\
v(x) &= \sum_{n=0}^{\infty} v_n(x),
\end{aligned}
\tag{96}
$$

where $u_n(x)$ and $v_n(x), n \geq 0$ are the components of $u(x)$ and $v(x)$ that will be computed in a recursive manner.

Substituting (96) into (95) gives

$$
\begin{aligned}
\sum_{n=0}^{\infty} u_n(x) &= x + \sin x - \frac{\pi^2}{4} - \frac{\pi}{2} x \\
&\quad + \int_0^{\frac{\pi}{2}} \left((1 + xt) \sum_{n=0}^{\infty} u_n(t) + (1 - xt) \sum_{n=0}^{\infty} v_n(t) \right) dt, \\
\sum_{n=0}^{\infty} v_n(x) &= x - \cos x - \frac{\pi^2}{4} + \frac{\pi}{2} x \\
&\quad + \int_0^{\frac{\pi}{2}} \left((1 - xt) \sum_{n=0}^{\infty} u_n(t) + (1 + xt) \sum_{n=0}^{\infty} v_n(t) \right) dt.
\end{aligned}
\tag{97}
$$

The Adomian decomposition method admits the use of the recurrence relation

$$
\begin{aligned}
u_0(x) &= x + \sin x - \frac{\pi^2}{4} - \frac{\pi}{2} x, \\
v_0(x) &= x - \cos x - \frac{\pi^2}{4} + \frac{\pi}{2} x, \\
u_1(x) &= \int_0^{\frac{\pi}{2}} \left((1 + xt) u_0(t) + (1 - xt) v_0(t) \right) dt = \frac{\pi^2}{4} + \frac{\pi}{2} x - \frac{\pi^3}{4} + \frac{\pi^4}{24} x, \\
v_1(x) &= \int_0^{\frac{\pi}{2}} \left((1 - xt) u_0(t) + (1 + xt) v_0(t) \right) dt = \frac{\pi^2}{4} - \frac{\pi}{2} x - \frac{\pi^3}{4} - \frac{\pi^4}{24} x.
\end{aligned}
\tag{98}
$$

By canceling the noise terms from $u_0(x)$ and from $v_0(x)$ we obtain the exact solutions

$$
(u(x), v(x)) = (x + \sin x, x - \cos x).
\tag{99}
$$

Recall that the remaining terms of $u_1(x)$ and $v_1(x)$ will be cancelled with the noise terms that will appear when evaluating the components $u_2(x)$ and $v_2(x)$. Generally, all noise terms will vanish in the limit.

In the next example, to review the methods presented earlier, we will apply the direct computation method. Note that we can also use the modified decomposition method.

Example 2. Solve the following system of Fredholm integral equations by using the direct computation method

$$
\begin{aligned}
u(x) &= x^2 + \sin x - \frac{\pi^3}{12} x + \int_0^{\frac{\pi}{2}} (xu(t) + xv(t)) \, dt, \\
v(x) &= x^2 - \cos x - 2x + \int_0^{\frac{\pi}{2}} (xu(t) - xv(t)) \, dt.
\end{aligned}
\tag{100}
$$

Following the analysis presented above, this system can be rewritten as

$$
\begin{aligned}
u(x) &= x^2 + \sin x + (\alpha - \frac{\pi^3}{12})x, \\
v(x) &= x^2 - \cos x + (\beta - 2)x,
\end{aligned}
\tag{101}
$$

where

$$
\begin{aligned}
\alpha &= \int_0^{\frac{\pi}{2}} (u(t) + v(t))\, dt, \\
\beta &= \int_0^{\frac{\pi}{2}} (u(t) - v(t))\, dt.
\end{aligned}
\tag{102}
$$

Substitute (101) into (102) and solving the resulting equations we find

$$
\alpha = \frac{\pi^3}{12}, \quad \beta = 2.
\tag{103}
$$

This in turn gives the exact solutions

$$
(u(x), v(x)) = (x^2 + \sin x, x^2 - \cos x).
\tag{104}
$$

9.5.2   Systems of Volterra Integral Equations

Systems of Volterra integral equations appear in many scientific applications such as Volterra population growth models and species propagation. Recall that Volterra integral equations are characterized by at least one variable limit of integration. Proceeding as before, we will concern ourselves for systems of Volterra integral equations of the second kind where only two unknown functions $u(x)$ and $v(x)$ are involved.

   The standard system of Volterra integral equations of the second kind, the unknown functions $u(x)$ and $v(x)$ appear inside and outside the integral sign of the form

$$
\begin{aligned}
u(x) &= f_1(x) + \int_0^x \left(K_1(x,t)u(t) + \tilde{K}_1(x,t)v(t) \right) dt, \\
v(x) &= f_2(x) + \int_0^x \left(K_2(x,t)u(t) + \tilde{K}_2(x,t)v(t) \right) dt,
\end{aligned}
\tag{105}
$$

$$\vdots$$

The kernels $K_1(x,t)$, $K_2(x)$, $\tilde{K}_1(x,t)$, $\tilde{K}_2(x,t)$, and the functions $f_1(x)$ and $f_2(x)$ are given real-valued functions.

   To avoid the cumbersome work that usually arise from the traditional methods, we will use the Adomian decomposition method and the modified decomposition method in this section.

Example 3. Use the modified decomposition method to solve the following system of Volterra integral equations of the second kind

$$
\begin{aligned}
u(x) &= x^2 - \frac{1}{12}x^5 + \int_0^x \left((x-t)^2 u(t) + (x-t)v(t) \right) \, dt, \\
v(x) &= x^3 + \frac{1}{60}x^5 + \int_0^x \left((x-t)^2 u(t) - (x-t)v(t) \right) \, dt.
\end{aligned}
\tag{106}
$$

The modified decomposition method suggests assumes that $u(x)$ and $v(x)$ be expressed by an infinite series of components

$$
\begin{aligned}
u(x) &= \sum_{n=0}^{\infty} u_n(x), \\
v(x) &= \sum_{n=0}^{\infty} v_n(x),
\end{aligned}
\tag{107}
$$

where $u_n(x)$ and $v_n(x), n \geq 0$ are the components of $u(x)$ and $v(x)$ that will be computed by a recursive manner.

Substituting (107) into (106) gives

$$
\begin{aligned}
\sum_{n=0}^{\infty} u_n(x) &= x^2 - \frac{1}{12}x^5 + \int_0^x \left((x-t)^2 \sum_{n=0}^{\infty} u_n(t) + (x-t) \sum_{n=0}^{\infty} v_n(t) \right) \, dt, \\
\sum_{n=0}^{\infty} v_n(x) &= x^3 + \frac{1}{60}x^5 + \int_0^x \left((x-t)^2 \sum_{n=0}^{\infty} u_n(t) 2(x-t) \sum_{n=0}^{\infty} v_n(t) \right) \, dt.
\end{aligned}
\tag{108}
$$

The modified method gives the recursive relation

$$
\begin{aligned}
u_0(x) &= x^2, \\
v_0(x) &= x^3, \\
u_1(x) &= -\frac{1}{12}x^5 + \int_0^x \left((x-t)^2 u_0(t) + (x-t) v_0(t) \right) \, dt = 0, \\
v_1(x) &= \frac{1}{60}x^5 + \int_0^x \left((x-t)^2 u_0(t) - (x-t) v_0(t) \right) \, dt = 0.
\end{aligned}
\tag{109}
$$

This in turn gives

$$
(u(x), v(x)) = (x^2, x^3).
\tag{110}
$$

Example 4. Use the modified decomposition method to solve the following system of Volterra integral equations of the second kind

$$
\begin{aligned}
u(x) &= e^x + x + \int_0^x \left(e^{-t} u(t) - 2e^t v(t) \right) \, dt, \\
v(x) &= e^{-x} - x + \int_0^x \left(2e^{-t} u(t) - e^t v(t) \right) \, dt.
\end{aligned}
\tag{111}
$$

Proceeding as before, the modified decomposition method gives the recursive relation

$$
\begin{aligned}
u_0(x) &= e^x, \\
v_0(x) &= e^{-x}, \\
u_1(x) &= x + \int_0^x \left(e^{-t} u_0(t) - 2e^t v_0(t) \right) dt = 0, \\
v_1(x) &= -x + \int_0^x \left(2e^{-t} u_0(t) - e^t v_0(t) \right) dt = 0.
\end{aligned}
\tag{112}
$$

This in turn gives

$$
(u(x), v(x)) = (e^x, e^{-x}).
\tag{113}
$$

9.6 Numerical Treatment of Fredholm Integral Equations

Fredholm and Volterra integral equations have several applications in physics, chemistry, biology, and engineering. For concrete integral equations, analytical solutions are hard to find. Due to this fact, several numerical methods have been developed for finding numerical solutions of integral equations. Examples of these methods include Chebyshev polynomials, Taylor series method, the radial basis functions, Bernstein's method, the wavelet method,, the homotopy perturbation method, approximation, the Toeplitz matrix method, and other methods. The need to apply numerical methods is necessary in order to have a scientific platform to be used for numerical purposes.

As stated before, the two newly developed methods, namely the Adomian decomposition method and the variational iteration method provide convergent series of the solution of any integral equation. For concrete problems, we may face some difficulties that need cumbersome work to obtain many terms of the series solution. In this case, we often determine few terms of the series solution, either by computing finite number of components, if Adomian method is used, or by evaluating few approximations, if the variational iteration method is used. In this case we use the obtained series for numerical purposes, where it was found that the obtained truncated series gives accuracy of higher level. In the following examples, we will use the Adomian method for numerical treatment of the Fredholm integral equations of the second kind. The accuracy level will be tested by showing the errors between the exact solution and the truncated series.

Example 1. Use the Adomian decomposition method to find ϕ_4 of the numerical solution and the errors between the exact solution and ϕ_4 of the

Fredholm integral equation of the second kind

$$u(x) = \frac{1}{16} \cos x(12 - 4\pi x + \pi^2) + \cos x \int_0^{\frac{\pi}{2}} (x - t)u^2(t)\, dt. \qquad (114)$$

The Adomian method is well known now, hence we skip details. The Adomian method admits the use of the recurrence relation

$$\begin{aligned}
u_0(x) &= \frac{1}{16} \cos x(12 - 4\pi x + \pi^2), \\
u_{k+1}(x) &= \cos x \int_0^{\frac{\pi}{2}} (x - t)A_k(t)\, dt, \, k \geq 0,
\end{aligned} \qquad (115)$$

where $A_k(x)$ are the Adomian polynomial for the nonlinear term $u^2(x)$. By evaluating the first four components only, and using the fourth-stage solution approximant

$$\phi_4(x) = \sum_{i=0}^{3} u_i(x), \qquad (116)$$

and by using the Taylor series for the approximant ϕ_4 we find that

$$\begin{aligned}
\phi_4 =\ & 0.9529469970 + 0.02954436161\, x \\
& - 0.4764734985\, x^2 - 0.01477218072\, x^3 \\
& + 0.03970612488\, x^4 + 0.001231015071\, x^5 \\
& - 0.001323537497\, x^6 - 0.00004103383560\, x^7 \\
& + 0.00002363459814\, x^8 + 0.0000007327470680\, x^9 \\
& - 0.0000002626066460\, x^{10} - 0.000000008141634054\, x^{11} \\
& + 0.000000001989444288\, x^{12} + O(x^{13}).
\end{aligned} \qquad (117)$$

Note that the exact solution $u(x) = \cos x$, therefore we list the following table of errors.

Table 1

x	Error $= |\cos(x) - \phi_4|$	x	Error $= |\cos(x) - \phi_4|$
0.0	4.705300334e-2	0.7	2.017038004e-2
0.1	4.387825847e-2	0.8	1.631513883e-2
0.2	4.032398704e-2	0.9	1.272005321e-2
0.3	3.648400803e-2	1.0	9.45995906e-3
0.4	3.245382203e-2	1.1	6.60173118e-3
0.5	2.832908706e-2	1.2	4.20326626e-3
0.6	2.420411131e-2	$\frac{\pi}{2}$	0.00000000000

This clearly shows that we obtained a series of higher accuracy level by using four components only.

Table 1 above shows the error between the exact solution and the approximation ϕ_4.

Example 2. Use the Adomian decomposition method to find the third-stage solution ϕ_3 of the numerical solution and the errors between the exact solution and ϕ_3 of the Fredholm integral equation of the second kind

$$u(x) = (2x - x\ln 2 + 1)\ln 2 - x - \frac{5}{8} + \ln(1+x) + \frac{1}{2}\int_0^1 (x-t)u^2(t)\, dt. \quad (118)$$

Proceeding as in Example 1, we set the recurrence relation

$$
\begin{aligned}
u_0(x) &= (2x - x\ln 2 + 1)\ln 2 - x - \frac{5}{8} + \ln(1+x), \\
u_{k+1}(x) &= \cos x \int_0^1 (x-t)A_k(t)\, dt, k \geq 0,
\end{aligned}
\quad (119)
$$

where $A_k(x)$ are the Adomian polynomial for the nonlinear term $u^2(x)$. By evaluating the first three components only, and using the approximant

$$\phi_3(x) = \sum_{i=0}^{2} u_i(x), \quad (120)$$

and by using the Taylor series for the approximant ϕ_3 we find that

$$
\begin{aligned}
\phi_3 = \ & -0.00053678031 + 1.000338540\, x - 0.5000000000\, x^2 \\
& + 0.3333333333\, x^3 - 0.2500000000\, x^4 + 0.2000000000\, x^5 \\
& - 0.1666666667\, x^6 + 0.1428571429\, x^7 - 0.1250000000\, x^8 \\
& + 0.1111111111\, x^9 - 0.1000000000\, x^{10} + 0.09090909091\, x^{11} \\
& - 0.08333333333\, x^{12} + O(x^{13}).
\end{aligned}
\quad (121)
$$

Note that the exact solution is $u(x) = \ln(1+x)$.

Table 2

x	Error $= \lvert \ln(1+x) - \phi_3 \rvert$	x	Error $= \lvert \ln(1+x) - \phi_3 \rvert$
0.0	5.36780e-4	0.6	3.33660e-4
0.1	5.02926e-4	0.7	2.99803e-4
0.2	4.69072e-4	0.8	2.65950e-4
0.3	4.35219e-4	0.9	2.32096e-4
0.4	4.01365e-4	1.0	1.98243e-4
0.5	3.67515e-4		

Table 2 above shows the error between the exact solution and the approximation ϕ_3.

Example 3. Use the Adomian decomposition method to find the approximation ϕ_{12} and ψ_{12} of the numerical solution and the errors between the exact solutions and ϕ_{12} and ψ_{12} of the system of Fredholm integral equations of the second kind

$$
\begin{aligned}
u(x) &= \frac{1}{18}x + \frac{17}{36} + \int_0^1 \frac{1}{3}(x+t)(u(t)+v(t))\,dt, \\
v(x) &= x^2 - \frac{19}{12}x + 1 + \int_0^1 xt(u(t)+v(t))\,dt.
\end{aligned}
\tag{122}
$$

Proceeding as in Example 1, we set the recurrence relations

$$
\begin{aligned}
u_0(x) &= \frac{1}{18}x + \frac{17}{36}, \\
u_{k+1}(x) &= \int_0^1 \frac{1}{3}(x+t)(u_k(t)+v_k(t))\,dt, \ k \geq 0,
\end{aligned}
\tag{123}
$$

and

$$
\begin{aligned}
v_0(x) &= x^2 - \frac{19}{12}x + 1, \\
v_{k+1}(x) &= \int_0^1 xt(u_k(t)+v_k(t))\,dt, \ k \geq 0.
\end{aligned}
\tag{124}
$$

This in turn gives

$$
\begin{aligned}
u_0(x) &= \frac{1}{18}x + \frac{17}{36}, \\
v_0(x) &= x^2 - \frac{19}{12}x + 1, \\
u_1(x) &= \frac{25}{72}x + \frac{103}{648}, \\
v_1(x) &= \frac{103}{216}x, \\
u_2(x) &= \frac{185}{972}x + \frac{17}{144}, \\
v_2(x) &= \frac{17}{58}x,
\end{aligned}
\tag{125}
$$

and so on for other components. By evaluating the first twelve components for $u(x)$ and $v(x)$, the approximations

$$\phi_{12}(x) = \sum_{i=0}^{11} u_i(x) = 0.981305x + 0.988512,$$
$$\psi_{12}(x) = \sum_{i=0}^{11} v_i(x) = 1 - 0.034504x + x^2, \tag{126}$$

are readily obtained.

In this example, we found it necessary to determine twelve components for the solutions $u(x)$ and $v(x)$ to enhance the accuracy level. To minimize the errors between the exact solutions and the approximations, we should evaluate more components.

Note that the exact solutions are given by

$$(u(x), v(x)) = 1 + x, 1 + x^2. \tag{127}$$

In view of this, we present the following table of errors.

Table 3

x	Error $= |(1 + x) - \phi_{12}|$	Error $= |(1 + x^2) - \psi_{12}|$
0.0	1.14880e-2	0.0
0.1	1.33575e-2	3.45040e-3
0.2	1.52270e-2	6.90080e-3
0.3	1.70965e-2	1.03512e-2
0.4	1.89660e-2	1.38016e-2
0.5	2.08355e-2	1.72520e-2
0.6	2.27050e-2	2.07024e-2
0.7	2.45745e-2	2.41528e-2
0.8	2.64440e-2	2.76032e-2
0.9	2.83135e-2	3.10536e-2
1.0	3.01830e-2	3.45040e-2

Table 3 above shows the error between the exact solutions and the approximations ϕ_{12} and ψ_{12}.

9.7 Numerical Treatment of Volterra Integral Equations

In a manner parallel to the analysis presented in the previous section, we will handle concrete problems of the Volterra integral equations, where some difficulties may arise. In this case, we prefer to handle the problem numerically, and we often determine few terms of the series solution, either by computing finite number of components, if Adomian method is used, or by evaluating few approximations, if the variational iteration method is used. In this case we use the obtained series for numerical purposes, where it was found that the obtained truncated series gives accuracy of higher level. It is worth noting that the accuracy level can be enhanced by determining more components, if ADM is used, or more approximations if VIM is used. In the following examples, we will use the Adomian method for numerical treatment of the Volterra integral equations of the second kind. The accuracy level will be tested by showing the errors between the exact solution and the truncated series.

Example 1. Use the Adomian decomposition method to find ϕ_4 of the numerical solution and the errors between the exact solution and ϕ_3 of the Volterra integral equation of the second kind

$$u(x) = \frac{1}{4} \cos x(3 - x^2 + \cos^2 x) + \int_0^x (x - t) \cos x\, u^2(t)\, dt. \tag{128}$$

The Adomian method is well known now, hence we skip the details. The Adomian method admits the use of the recurrence relation

$$\begin{aligned} u_0(x) &= \frac{1}{4} \cos x(3 - x^2 + \cos^2 x), \\ u_{k+1}(x) &= \int_0^x (x - t) \cos x A_k(t)\, dt, k \geq 0, \end{aligned} \tag{129}$$

where $A_k(x)$ are the Adomian polynomial for the nonlinear term $u^2(x)$. By evaluating the first four components only, and using the approximant

$$\phi_4(x) = \sum_{i=0}^3 u_i(x), \tag{130}$$

and by using the Taylor series for the approximant ϕ_4 we find that

$$\begin{aligned} \phi_4 &= 1 - 0.5x^2 + 0.04166666667x^4 - 0.001388888889ex^6 \\ &\quad - 0.001959325397x^8 + 0.003405809083x^{10} - 0.002818493717x^{12} \\ &\quad + 0.001541612862x^{14} - 0.0006376478536x^{16} + O(x^{17}). \end{aligned} \tag{131}$$

As stated before, we can minimize the errors between the exact solution and the approximant, by evaluating more components of the solution. Note that the exact solution is given by

$$u(x) = \cos x. \tag{132}$$

Based on this, we list the following table of errors.

Table 4

x	Error $= \lvert \cos(x) - \phi_4 \rvert$	x	Error $= \lvert \cos(x) - \phi_4 \rvert$
0.0	1.080430170e-8	0.7	4.85318893e-5
0.1	6.996036355e-9	0.8	1.07732322e-4
0.2	1.18811309e-8	0.9	2.01940664e-4
0.3	1.11844732e-7	1.0	3.26754224e-4
0.4	9.9126378e-7	1.1	4.61429093e-4
0.5	5.0337891e-6	1.2	5.68707641e-4
0.6	1.78202512e-5	$\frac{\pi}{2}$	0.00000000000

Table 4 above shows the error between the exact solution and the approximation ϕ_4.

This clearly shows we obtained a series of higher accuracy level by using few components only. The accuracy level can be enhanced by the determination of new components.

Example 2. Use the Adomian decomposition method to find ϕ_3 of the numerical solution and the errors between the exact solution and ϕ_3 of the Volterra integral equation of the second kind

$$u(x) = (3 + 2x)\ln(1 + x) - (1 + x)\ln^2(1 + x) - 2x + \int_0^x u^2(t)\, dt. \tag{133}$$

Proceeding as in Example 1, we set the recurrence relation

$$\begin{aligned} u_0(x) &= (3 + 2x)\ln(1 + x) - (1 + x)\ln^2(1 + x) - 2x, \\ u_{k+1}(x) &= \int_0^x A_k(t)\, dt,\ k \geq 0, \end{aligned} \tag{134}$$

where $A_k(x)$ are the Adomian polynomial for the nonlinear term $u^2(x)$. By evaluating the first three components only, and using the approximant

$$\phi_3(x) = \sum_{i=0}^{2} u_i(x), \tag{135}$$

and by using the Taylor series for the approximant ϕ_3 we find that

$$
\begin{aligned}
\phi_3 \;=\;& x - .5x^2 + .3333333333x^3 - .25x^4 + .2x^5 - .1666666667x^6 \\
&+ 0.08888888889x^7 - 0.05277777778x^8 + 0.05205026455ex^9 \\
&- 0.05548280423ex^{10} + 0.05625761584x^{11} - 0.05533472957x^{12} \\
&+ O(x^{13}).
\end{aligned}
$$

(136)

Note that the exact solution is $u(x) = \ln(1 + x)$, hence, we present the following table of errors

Table 5

x	Error $= |\ln(1+x) - \phi_3|$	x	Error $= |\ln(1+x) - \phi_3|$
0.1	4.7e-9	0.6	7.08677e-4
0.2	5.315e-7	0.7	1.84631e-3
0.3	8.016e-6	0.8	4.16473e-3
0.4	5.300296e-5	0.9	8.41388e-3
0.5	2.23463816e-4	1.0	1.557791e-2

Table 5 above shows the error between the exact solution and the approximation ϕ_3.

Recall that the more components we determine the more accuracy level we achieve.

Example 3. Use the Adomian decomposition method to find ϕ_4 of the numerical solution and the errors between the exact solution and ϕ_4 of the Volterra integral equation of the first kind

$$
\sin x = \int_0^x \cos(x - t)u(t)\,dt. \tag{137}
$$

We first convert this equation from first kind to a second kind by using Leibniz rule. This in turn gives

$$
u(x) = \cos x + \int_0^x \sin(x - t)u(t)\,dt. \tag{138}
$$

We next set the recurrence relation

$$
\begin{aligned}
u_0(x) &= \cos x, \\
u_{k+1}(x) &= \int_0^x u_k(t)\,dt, \; k \ge 0.
\end{aligned} \tag{139}
$$

By evaluating the first four components only, and using the approximant

$$\phi_4(x) = \sum_{i=0}^{3} u_i(x), \tag{140}$$

we obtain

$$\phi_4(x) = \cos(x) + \frac{11}{16}x \sin x - \frac{3}{16}x^2 \cos x - \frac{1}{48}x^3 \sin x. \tag{141}$$

Note that the exact solution is $u(x) = 1$, hence, we present the following table of errors

Table 6

x	Error $= |\ln(1+x) - \phi_4|$	x	Error $= |\ln(1+x) - \phi_4|$
0.1	3.2e-11	0.6	4.1e-7
0.2	8.0e-11	0.7	1.4e-6
0.3	1.68e-9	0.8	4.0e-6
0.4	1.61e-8	0.9	1.0e-5
0.5	9.58e-8	1.0	2.4e-5

Table 6 above shows the error between the exact solution and the approximation ϕ_4.

Recall that the more components we determine the more accuracy level we achieve.

Appendix A

Table of Indefinite Integrals

I. Basic Forms:

1. $\int x^n \, dx = \dfrac{1}{n+1} x^{n+1} + C, n \neq -1.$

2. $\int \dfrac{1}{x} \, dx = \ln|x| + C.$

3. $\int e^x \, dx = e^x + C.$

4. $\int \dfrac{1}{1+x^2} \, dx = \tan^{-1} x + C.$

5. $\int \dfrac{1}{\sqrt{1-x^2}} \, dx = \sin^{-1} x + C.$

6. $\int \cos x \, dx = \sin x + C.$

7. $\int \sin x \, dx = -\cos x + C.$

8. $\int \tan x \, dx = -\ln|\cos x| + C.$

9. $\int \tan x \, \sec x \, dx = \sec x + C.$

10. $\int \sec^2 x \, dx = \tan x + C.$

II. Trigonometric Forms:

1. $\displaystyle\int \sin^2 x\,dx = \frac{1}{2}x - \frac{1}{4}\sin 2x + C.$

2. $\displaystyle\int \cos^2 x\,dx = \frac{1}{2}x + \frac{1}{4}\sin 2x + C.$

3. $\displaystyle\int \sin^3 x\,dx = -\frac{1}{3}\cos x\left(2 + \sin^2 x\right) + C.$

4. $\displaystyle\int \cos^3 x\,dx = \frac{1}{3}\sin x\left(2 + \cos^2 x\right) + C.$

5. $\displaystyle\int \tan^2 x\,dx = \tan x - x + C.$

6. $\displaystyle\int \cot^2 x\,dx = -\cot x - x + C.$

7. $\displaystyle\int x \sin x\,dx = \sin x - x\cos x + C.$

8. $\displaystyle\int x \cos x\,dx = \cos x + x\sin x + C.$

9. $\displaystyle\int x^2 \sin x\,dx = 2x\sin x - \left(x^2 - 2\right)\cos x + C.$

10. $\displaystyle\int x^2 \cos x\,dx = 2x\cos x + \left(x^2 - 2\right)\sin x + C.$

III. Inverse Trigonometric Forms:

1. $\displaystyle\int \sin^{-1} x\,dx = x\sin^{-1} x + \sqrt{1 - x^2} + C.$

2. $\displaystyle\int \cos^{-1} x\,dx = x\cos^{-1} x - \sqrt{1 - x^2} + C.$

3. $\displaystyle\int \tan^{-1} x\,dx = x\tan^{-1} x - \frac{1}{2}\ln(1 + x^2) + C.$

4. $\displaystyle\int x\sin^{-1} x\,dx = \frac{1}{4}[(2x^2 - 1)\sin^{-1} x + x\sqrt{1 - x^2}] + C.$

5. $\displaystyle\int x\cos^{-1} x\,dx = \frac{1}{4}[(2x^2 - 1)\cos^{-1} x - x\sqrt{1 - x^2}] + C.$

6. $\displaystyle\int x\tan^{-1} x\,dx = \frac{1}{2}[(x^2 + 1)\tan^{-1} x - x] + C.$

7. $\displaystyle\int \sec^{-1} x\,dx = x\sec^{-1} x - \ln(x + \sqrt{x^2 - 1}) + C.$

8. $\int x\sec^{-1}x\,dx = \frac{1}{2}[x^2\sec^{-1}x - \sqrt{x^2-1}] + C.$

IV. Exponential and Logarithmic Functions Forms:

1. $\int e^{ax}\,dx = \frac{1}{a}e^{ax} + C.$

2. $\int xe^{ax}\,dx = \frac{1}{a^2}(ax-1)e^{ax} + C.$

3. $\int x^2 e^{ax}\,dx = \frac{1}{a^3}(a^2x^2 - 2ax + 2)e^{ax} + C.$

4. $\int x^3 e^{ax}\,dx = \frac{1}{a^4}(a^3x^3 - 3a^2x^2 + 6ax - 6)e^{ax} + C.$

5. $\int e^x \sin x\,dx = \frac{1}{2}(\sin x - \cos x)e^x + C.$

6. $\int e^x \cos x\,dx = \frac{1}{2}(\sin x + \cos x)e^x + C.$

7. $\int \ln x\,dx = x\ln x - x + C.$

8. $\int x\ln x\,dx = \frac{1}{2}x^2(\ln x - \frac{1}{2}) + C.$

V. Hyperbolic Functions Forms:

1. $\int \sinh x\,dx = \cosh x + C.$

2. $\int \cosh x\,dx = \sinh x + C.$

3. $\int x\sinh x\,dx = x\cosh x - \sinh x + C.$

4. $\int x\cosh x\,dx = x\sinh x - \cosh x + C.$

5. $\int \sinh^2 x\,dx = \frac{1}{2}(\sinh x\cosh x - x) + C.$

6. $\int \cosh^2 x\,dx = \frac{1}{2}(\sinh x\cosh x + x) + C.$

Appendix B

Integrals Involving Irrational Algebraic Functions

I. **Integrals Involving** $\frac{t^n}{\sqrt{x-t}}$, n **is an Integer,** $n \geq 0$:

1. $\displaystyle\int_0^x \frac{1}{\sqrt{x-t}}\,dt = 2\sqrt{x}.$

2. $\displaystyle\int_0^x \frac{t}{\sqrt{x-t}}\,dt = \frac{4}{3}x^{\frac{3}{2}}.$

3. $\displaystyle\int_0^x \frac{t^2}{\sqrt{x-t}}\,dt = \frac{16}{15}x^{\frac{5}{2}}.$

4. $\displaystyle\int_0^x \frac{t^3}{\sqrt{x-t}}\,dt = \frac{32}{35}x^{\frac{7}{2}}.$

5. $\displaystyle\int_0^x \frac{t^4}{\sqrt{x-t}}\,dt = \frac{256}{315}x^{\frac{9}{2}}.$

6. $\displaystyle\int_0^x \frac{t^5}{\sqrt{x-t}}\,dt = \frac{512}{693}x^{\frac{11}{2}}.$

7. $\displaystyle\int_0^x \frac{t^6}{\sqrt{x-t}}\,dt = \frac{2048}{3003}x^{\frac{13}{2}}.$

II. **Integrals Involving** $\frac{t^{\frac{n}{2}}}{\sqrt{x-t}}$, n **is an Odd Integer,** $n \geq 1$:

1. $\displaystyle\int_0^x \frac{t^{\frac{1}{2}}}{\sqrt{x-t}}\,dt = \frac{1}{2}\pi x.$

287

2. $\int_0^x \dfrac{t^{\frac{3}{2}}}{\sqrt{x-t}}\,dt = \dfrac{3}{8}\pi x^2.$

3. $\int_0^x \dfrac{t^{\frac{5}{2}}}{\sqrt{x-t}}\,dt = \dfrac{5}{16}\pi x^3.$

4. $\int_0^x \dfrac{t^{\frac{7}{2}}}{\sqrt{x-t}}\,dt = \dfrac{35}{128}\pi x^4.$

5. $\int_0^x \dfrac{t^{\frac{9}{2}}}{\sqrt{x-t}}\,dt = \dfrac{63}{256}\pi x^5.$

6. $\int_0^x \dfrac{t^{\frac{11}{2}}}{\sqrt{x-t}}\,dt = \dfrac{231}{1024}\pi x^6.$

7. $\int_0^x \dfrac{t^{\frac{13}{2}}}{\sqrt{x-t}}\,dt = \dfrac{429}{2048}\pi x^7.$

8. $\int_0^x \dfrac{t^{\frac{15}{2}}}{\sqrt{x-t}}\,dt = \dfrac{6435}{32768}\pi x^8.$

9. $\int_0^x \dfrac{t^{\frac{17}{2}}}{\sqrt{x-t}}\,dt = \dfrac{12155}{65536}\pi x^9.$

10. $\int_0^x \dfrac{t^{\frac{n}{2}}}{\sqrt{x-t}}\,dt = \dfrac{\Gamma(\frac{n+2}{2})}{\Gamma(\frac{n+3}{2})}\sqrt{\pi}x^{\frac{n+1}{2}}.$

Appendix C

Series

I. Exponential Series:

1. $e^x = 1 + x + \dfrac{x^2}{2!} + \dfrac{x^3}{3!} + \dfrac{x^4}{4!} + \cdots$.

2. $e^x = 1 - x + \dfrac{x^2}{2!} - \dfrac{x^3}{3!} + \dfrac{x^4}{4!} + \cdots$.

3. $e^{-x^2} = 1 - x^2 + \dfrac{x^4}{2!} - \dfrac{x^6}{3!} + \cdots$.

4. $a^x = 1 + x \ln a + \dfrac{1}{2!}(x \ln a)^2 + \dfrac{1}{3!}(x \ln a)^3 + \cdots, a > 0$.

II. Trigonometric Series:

1. $\sin x = x - \dfrac{x^3}{3!} + \dfrac{x^5}{5!} - \dfrac{x^7}{7!} + \cdots$.

2. $\cos x = 1 - \dfrac{x^2}{2!} + \dfrac{x^4}{4!} - \dfrac{x^6}{6!} + \cdots$.

3. $\sin x + \cos x = (1 + x) - \left(\dfrac{x^2}{2!} + \dfrac{x^3}{3!}\right) + \left(\dfrac{x^4}{4!} + \dfrac{x^5}{5!}\right) - \cdots$.

III. Hyperbolic Functions Series:

1. $\sinh x = x + \dfrac{x^3}{3!} + \dfrac{x^5}{5!} + \dfrac{x^7}{7!} + \cdots.$

2. $\cosh x = 1 + \dfrac{x^2}{2!} + \dfrac{x^4}{4!} + \dfrac{x^6}{6!} + \cdots.$

Appendix D

The Error and the Gamma Functions

I. **The Error Function:**

The *error function* erf (x) is defined by:

1. $\text{erf}(x) = \int_0^x e^{-u^2} du.$

2. $\text{erf}(x) = \dfrac{2}{\sqrt{\pi}} \left(x - \dfrac{x^3}{3} + \dfrac{x^5}{5 \cdot 2!} - \dfrac{x^7}{7 \cdot 3!} + \cdots \right).$

The *complementary error function* erfc(x) is defined by:

3. $\text{erfc}(x) = \dfrac{2}{\sqrt{\pi}} \int_x^\infty e^{-u^2} du.$

4. $\text{erf}(x) + \text{erfc}(x) = 1.$

5. $\text{erfc}(x) = 1 - \dfrac{2}{\sqrt{\pi}} \left(x - \dfrac{x^3}{3} + \dfrac{x^5}{5 \cdot 2!} - \dfrac{x^7}{7 \cdot 3!} + \cdots \right).$

II. **The Gamma Function $\Gamma(x)$:**

1. $\Gamma(x) = \int_0^\infty t^{x-1} e^{-t} dt.$

2. $\Gamma(x+1) = x\Gamma(x).$

3. $\Gamma(1) = 1, \Gamma(n+1) = n!, \quad n$ is an integer.

4. $\Gamma(x)\Gamma(1-x) = \dfrac{\pi}{\sin \pi x}$.

5. $\Gamma(1/2) = \sqrt{\pi}$.

Answers to Exercises

Exercises   1.2

1. Fredholm, linear, nonhomogeneous
2. Volterra, linear, nonhomogeneous
3. Volterra, nonlinear, nonhomogeneous
4. Fredholm, linear, homogeneous
5. Fredholm, linear, nonhomogeneous
6. Fredholm, nonlinear, nonhomogeneous
7. Fredholm, nonlinear, nonhomogeneous
8. Fredholm, linear, nonhomogeneous
9. Volterra, nonlinear, nonhomogeneous
10. Volterra, linear, nonhomogeneous
11. Volterra integro-differential equation, nonlinear
12. Fredholm integro-differential equation, linear
13. Volterra integro-differential equation, nonlinear
14. Fredholm integro-differential equation, linear
15. Volterra integro-differential equation, linear

16. $u(x) = 1 + \int_0^x 4u(t)dt$

17. $u(x) = 1 + \int_0^x 3t^2 u(t)dt$

18. $u(x) = 4 + \int_0^x u^2(t)dt$

19. $u^{'}(x) = 1 + \int_0^x 4tu^2(t)dt, \; u(0) = 2$

20. $u^{'}(x) = 1 + \int_0^x 2tu(t)dt, \; u(0) = 0$

21.   Volterra–Fredholm integral equation, nonlinear, nonhomogeneous
22.   Volterra–Fredholm integro-differential equation, linear, nonhomoge-

neous

23. Volterra–Fredholm integro-differential equation, nonlinear, nonhomogeneous

24. Volterra (singular) integral equation, nonlinear, nonhomogeneous

Exercises   1.3

11. $f(x) = x^2$,   12. $f(x) = 1 + x$,   13. $\alpha = 1$,   14. $f(x) = \sin x$

Exercises   1.4

1. $\displaystyle\int_0^x 3(x-t)^2 u(t)dt$

2. $2xe^{x^3} - e^{x^2} + \displaystyle\int_x^{x^2} te^{xt}dt$

3. $\displaystyle\int_0^x 4(x-t)^3 u(t)dt$

4. $4\sin 5x - \sin 2x + \displaystyle\int_x^{4x} \cos(x+t)dt$

5. $u'''(x) = 2u(x),\ u(0) = u'(0) = 1, u''(0) = 0$

6. $u''(x) + u(x) = e^x,\ u(0) = u'(0) = 1$

7. $u''(x) - u(x) = 0,\ u(0) = 0, u'(0) = 1$

8. $u''(x) - u(x) = \cos x,\ u(0) = -1, u'(0) = 1$

9. $u''(x) - u'(x) - 2u(x) = 10,\ u(0) = 2, u'(0) = 5$

10. $u''(x) - 5u'(x) + 6u(x) = 0,\ u(0) = -5, u'(0) = -19$

11. $u'(x) + u(x) = \sec^2 x,\ u(0) = 0$

12. $u'''(x) - 3u''(x) - 6u'(x) + 5u(x) = 0,$
$u(0) = 1, u'(0) = 4, u''(0) = 23$

13. $u'''(x) - 4u(x) = 24x,$
$u(0) = u'(0) = 0, u''(0) = 2$

14. $u^{iv}(x) - u(x) = 0,$
$u(0) = u'(0) = 0, u''(0) = 2, u'''(0) = 0$

Exercises   1.5

1. $u(x) = -1 - \displaystyle\int_0^x u(t)dt,\ where\ y'(x) = u(x)$

2. $u(x) = x + \displaystyle\int_0^x u(t)dt,\ where\ y'(x) = u(x)$

3. $u(x) = \sec^2(x) - \displaystyle\int_0^x u(t)dt,\ where\ y'(x) = u(x)$

In problems 4–10, set $y''(x) = u(x)$

4. $u(x) = -1 - \displaystyle\int_0^x (x-t)u(t)dt,$

5. $u(x) = 1 + x + \int_0^x (x-t)u(t)dt$,

6. $u(x) = -11 - 6x - \int_0^x [5 + 6(x-t)]\, u(t)dt$

7. $u(x) = -1 - \int_1^x u(t)dt$

8. $u(x) = -1 + 4x + \int_0^x [2(x-t) - 1]\, u(t)dt$

9. $u(x) = \sin x - \int_0^x (x-t)u(t)dt$

10. $u(x) = x - \sin x + xe^x - e^x - \int_0^x [(x-t)e^x - \sin x]\, u(t)dt$

In problems 11–15, set $y'''(x) = u(x)$

11. $u(x) = 2x - x^2 + \int_0^x \left[1 + (x-t) - \frac{1}{2}(x-t)^2\right] u(t)dt$

12. $u(x) = -3x - 4 \int_0^x (x-t)u(t)dt$

13. $u(x) = 2 + x - \frac{1}{2}x^2 - \frac{1}{6}x^3 - \int_0^x \left[2(x-t) + \frac{1}{6}(x-t)^3\right] u(t)dt$

14. $u(x) = 1 - \frac{1}{2}x^2 + \frac{1}{3!}\int_0^x (x-t)^3 u(t)dt$

15. $u(x) = 2e^x - 1 - x - \int_0^x (x-t)u(t)dt$

Exercises   1.6

1. $u(x) = \sin x + \int_0^1 K(x,t)u(t)\, dt$,
where the kernel $K(x,t)$ is defined by
$$K(x,t) = \begin{cases} 4t(1-x) & 0 \le t \le x \\ 4x(1-t) & x \le t \le 1 \end{cases}$$

2. $u(x) = 1 + \int_0^1 K(x,t)u(t)dt$,
where the kernel $K(x,t)$ is defined by
$$K(x,t) = \begin{cases} 2xt(1-x) & 0 \le t \le x \\ 2x^2(1-t) & x \le t \le 1 \end{cases}$$

3. $u(x) = (2x - 1) + \int_0^1 K(x,t)u(t)dt$,
where the kernel $K(x,t)$ is defined by
$$K(x,t) = \begin{cases} t(1-x) & 0 \le t \le x \\ x(1-t) & x \le t \le 1 \end{cases}$$

4. $u(x) = (x-1) + \int_0^1 K(x,t)u(t)dt,$

where the kernel $K(x,t)$ is defined by

$$K(x,t) = \begin{cases} t & 0 \le t \le x \\ x & x \le t \le 1 \end{cases}$$

Exercises 1.7

1. $f(x) = e^{2x}$ 2. $f(x) = e^{-3x}$

3. $f(x) = e^x - 1$ 4. $f(x) = \cos(2x)$

5. $f(x) = \sin(3x)$ 6. $f(x) = \sinh(2x)$

7. $f(x) = \cosh(2x)$ 8. $f(x) = \cosh(3x) - 1$

9. $f(x) = 1 + \cos(2x)$ 10. $f(x) = 1 + \sin x$

Exercises 2.2

1. $u(x) = 4x$ 2. $u(x) = x^3$

3. $u(x) = x^2 + \dfrac{3}{8}x$ 4. $u(x) = 1 + e^x$

5. $u(x) = \sin x$ 6. $u(x) = \cos x$

7. $u(x) = \cos(4x)$ 8. $u(x) = \sinh x$

9. $u(x) = 2e^{2x}$ 10. $u(x) = \sec^2 x$

11. $u(x) = \sin x$ 12. $u(x) = \tan x$

13. $u(x) = \tan^{-1} x$ 14. $u(x) = \cosh x$

15. $u(x) = \dfrac{1}{1+x^2}$ 16. $u(x) = \dfrac{1}{\sqrt{1-x^2}}$

17. $u(x) = \dfrac{1}{1+x^2}$ 18. $u(x) = \cos^{-1} x$

19. $u(x) = x\tan^{-1} x$ 20. $u(x) = x\sin^{-1} x + 1$

21. $u(x) = \dfrac{\sin x}{1+\sin x}$ 22. $u(x) = \dfrac{\sin x}{1+\cos x}$

23. $u(x) = \dfrac{\sec^2 x}{1+\tan x}$ 24. $u(x) = 1 + \sin x$

25. $u(x) = 1 + \sin x + \cos x$ 26. $u(x) = x\sin x$

Exercises 2.3

1. $u(x) = x^3$ 2. $u(x) = e^x$

3. $u(x) = x$ 4. $u(x) = x^2 + x^4$

5. $u(x) = x + e^x$ 6. $u(x) = x + e^{-x}$

7. $u(x) = 1 + x + x^2$ 8. $u(x) = e^x$

Exercises 2.4

1. $u(x) = xe^x$
2. $u(x) = x^2 - 2x + 1$
3. $u(x) = x \sin x$
4. $u(x) = e^{2x}$
5. $u(x) = 1 + \sec^2 x$
6. $u(x) = \sin(2x)$
7. $u(x) = x^2 - \dfrac{5}{18}x - \dfrac{5}{36}$
8. $u(x) = \sin x + \cos x$
9. $u(x) = \sec x \tan x$
10. $u(x) = x^2$
11. $u(x) = \sin x$
12. $u(x) = 1 + \dfrac{1}{2}\ln x$
13. $u(x) = x^3$
14. $u(x) = 1 + \dfrac{\pi}{4}\sec^2 x$

Exercises 2.5

1. $u(x) = x$
2. $u(x) = x^3$
3. $u(x) = 4x$
4. $u(x) = 1 + 2x$
5. $u(x) = 2\sin x$
6. $u(x) = \sec^2 x$
7. $u(x) = \sec x \tan x$
8. $u(x) = \cosh x$
9. $u(x) = e^x$
10. $u(x) = \sin x$

Exercises 2.6

1. $u(x) = 2x$
2. $u(x) = 1 - \dfrac{\pi}{10}\cos x$
3. $u(x) = x + 1$
4. $u(x) = \sin x + \cos x$
5. $u(x) = x^2$
6. $u(x) = x^3$
7. $u(x) = \sin x + \cos x$
8. $u(x) = 1 + \dfrac{\pi}{2}\sin x$
9. $u(x) = 1 + \dfrac{\pi}{4}\sec^2 x$
10. $u(x) = 1 + \dfrac{\pi}{12}\sec x \tan x$

Exercises 2.8

1. $u(x) = A$, A is a constant
2. $u(x) = 2Ax$
3. $u(x) = Ax$, A is a constant
4. $u(x) = A\cos x$
5. $u_1(x) = \dfrac{2}{\pi}A(\sin x + \cos x)$ $u_2(x) = \dfrac{2}{\pi}A(\sin x - \cos x)$
6. $u_1(x) = u_2(x) = \dfrac{2}{\pi}(A\sin x + B\cos x)$
7. $u(x) = A\sec x$
8. $u(x) = A\sec^2 x$
9. $u(x) = \dfrac{2}{\pi - 2}A\sin^{-1} x$
10. $u(x) = \alpha\left(3 - \dfrac{3}{2}x\right)$

Exercises   2.9

1. $u(x) = x,$ 2. $u(x) = 3x$
3. $u(x) = e^{-x},$ 4. $u(x) = x^2$
5. $u(x) = \cos x$ 6. $u(x) = \sin x$

Exercises   3.2

1. $u(x) = 4x$ 2. $u(x) = 1 + 2x$
3. $u(x) = e^{-x}$ 4. $u(x) = \sinh x$
5. $u(x) = \sin(3x)$ 6. $u(x) = \cos(2x)$
7. $u(x) = \sin x + \cos x$ 8. $u(x) = \cos x - \sin x$
9. $u(x) = e^x$ 10. $u(x) = e^{-x}$
11. $u(x) = 2 \cosh x$ 12. $u(x) = 2e^x - 1$
13. $u(x) = 2 \cos x - 1$ 14. $u(x) = 2 \cosh x - 1$
15. $u(x) = \cos x$ 16. $u(x) = \sec^2 x$
17. $u(x) = \cosh x$ 18. $u(x) = \sinh x$
19. $u(x) = x^3$ 20. $u(x) = \sec x \tan x$
21. $u(x) = 8x$ 22. $u(x) = 8x^2$
23. $u(x) = \sec^2 x$ 24. $u(x) = 1 + x + x^2$
25. $u(x) = x \sin x$ 26. $u(x) = \cosh^2 x$

Exercises   3.3

1. $u(x) = e^x$ 2. $u(x) = \sinh x$
3. $u(x) = \sin(3x) + \cos x$ 4. $u(x) = \cos(2x)$
5. $u(x) = \cos x + \sin x$ 6. $u(x) = \cos x - \sin x$
7. $u(x) = \sin x$ 8. $u(x) = \cosh x$

Exercises   3.4

1. $u(x) = 2x + 3x^2$ 2. $u(x) = 1 + x + x^2$
3. $u(x) = \sin x + \cos x + \cos x$ 4. $u(x) = 1 + x$
5. $u(x) = -e^{-x} + \sin x$ 6. $u(x) = e^{-2x}$
7. $u(x) = e^x$ 8. $u(x) = \sinh x$
9. $u(x) = 2 \cos x - 1$ 10. $u(x) = \cos x - \sin x$
11. $u(x) = \sinh x$ 12. $u(x) = \sin x$

Exercises 3.5

1. $u(x) = e^{-3x}$

2. $u(x) = \cosh x$

3. $u(x) = \cos x - \sin x$

4. $u(x) = e^x - 1$

5. $u(x) = e^x$

6. $u(x) = \dfrac{1}{2}\left(\cos x + \cosh x\right)$

7. $u(x) = \dfrac{1}{2}\left(\sin x + \sinh x\right)$

8. $u(x) = 2\cosh x - 2$

9. $u(x) = x$

10. $u(x) = x - x^2$

Exercises 3.6

1. $u(x) = e^{-x}$

2. $u(x) = \cos(3x)$

3. $u(x) = e^{2x}$

4. $u(x) = e^{-\frac{1}{4}x}$

5. $u(x) = 2\cos x$

6. $u(x) = e^{-x^2}$

7. $u(x) = \sinh x$

8. $u(x) = \cos x$

9. $u(x) = \cos x + \sin x$

10. $u(x) = \cos x - \sin x$

11. $u(x) = 1 + e^x$

12. $u(x) = 1 - \sinh x$

Exercises 3.7

1. $u(x) = e^x - 1$

2. $u(x) = e^x - x - 1$

3. $u(x) = x - \sin x$

4. $u(x) = -x + \sinh x$

5. $u(x) = -1 + \cos x$

6. $u(x) = 1 - x$

7. $u(x) = e^{2x}$

8. $u(x) = 2 + e^x$

9. $u(x) = 1 + \cos x$

10. $u(x) = 1 - \sin x$

11. $u(x) = 1 + \cosh x$

12. $u(x) = -1 + \cosh x$

Exercises 3.9

1. $u(x) = 2x$

2. $u(x) = e^{-x}$

3. $u(x) = e^x$

4. $u(x) = \sinh x$

5. $u(x) = \cos x$

6. $u(x) = \sec^2 x$

Exercises 4.3

1. $u(x) = \dfrac{1}{6}(1 + x)$

2. $u(x) = \dfrac{1}{6} - \dfrac{1}{63}x^2$

3. $u(x) = \sin x$

4. $u(x) = x^2$

5. $u(x) = \sec^2 x$

6. $u(x) = 1 - 3x - 3x^2$

7. $u(x) = \sin x$

8. $u(x) = \sin x + \cos x$

Exercises   4.4

1. $u(x) = \cosh x$ 2. $u(x) = x$
3. $u(x) = xe^x$ 4. $u(x) = x \sin x$
5. $u(x) = \sin x$ 6. $u(x) = x^3 - x^2 + x - 1$
7. $u(x) = \sin x$ 8. $u(x) = \sin x + \cos x$
9. $u(x) = \cos x$ 10. $u(x) = \sin x - \cos x$

Exercises   4.5

1. $u(x) = x \sin x$ 2. $u(x) = x \cos x$
3. $u(x) = x + \sec^2 x$ 4. $u(x) = 1 + 2x - 6x^2$
5. $u(x) = e^x$ 6. $u(x) = 1 - x + x^2 - x^3$

Exercises   4.6

1. $u(x) = x \cos x$ 2. $u(x) = 1 - e^x$
3. $u(x) = \sin x - \cos x$ 4. $u(x) = 2x - 6x^2$
5. $u(x) = \dfrac{1}{2} \sin(2x)$

Exercises   5.3

1. $u(x) = x \cos x$ 2. $u(x) = 1 - e^x$
3. $u(x) = \sin x - \cos x$ 4. $u(x) = 2x - 4x^2$
5. $u(x) = 1 - \sinh x$

Exercises   5.4

1. $u(x) = x + e^x$ 2. $u(x) = 1 + \cos x$
3. $u(x) = 2e^x$ 4. $u(x) = \sin x + \cos x$
5. $u(x) = 1 - \sin x$

Exercises   5.5

1. $u(x) = xe^x$ 2. $u(x) = x + \sin x$
3. $u(x) = x - \cos x$ 4. $u(x) = x + \cosh x$
5. $u(x) = x + e^x$ 6. $u(x) = \cos x - \sin x$

Exercises   5.6

1. $u(x) = \cosh x$ 2. $u(x) = \sin x$
3. $u(x) = \sinh x$ 4. $u(x) = x + e^x$
5. $u(x) = \sin x + \cos x$ 6. $u(x) = e^x$

Exercises 5.7

1. $u(x) = \dfrac{1}{2}\cos x + \dfrac{1}{2}\sin x + \dfrac{1}{2}e^x$ 　　2. $u(x) = \sin x$

3. $u(x) = 1 + \sin x$ 　　　　　　　　　4. $u(x) = 1 + \cosh x$

5. $u(x) = 1 + 4x$ 　　　　　　　　　6. $u(x) = \dfrac{1}{4} + \sin x$

7. $u(x) = \dfrac{1}{4}e^x - \dfrac{3}{4}e^{-x} - \dfrac{1}{2}\cos x$

Exercises 5.8

1. $u(x) = \cos x$ 　　　　2. $u(x) = e^{-x}$
3. $u(x) = \sinh x$ 　　　　4. $u(x) = e^x$
5. $u(x) = x + \cos x$ 　　6. $u(x) = \cosh x$

Exercises 6.2

1. $u(x) = 2\sqrt{x} + \dfrac{1}{\sqrt{x}}$ 　　　　　　　2. $u(x) = \sqrt{x}\left(\dfrac{4}{3}x - 1\right)$

3. $u(x) = \dfrac{1}{\pi\sqrt{x}}\left(1 + 2x + \dfrac{8}{3}x^2\right)$ 　　4. $u(x) = x^{\frac{3}{2}}$

5. $u(x) = x$ 　　　　　　　　　　　　6. $u(x) = \dfrac{1}{2}x^2$

7. $u(x) = \dfrac{16}{5\pi}x^{\frac{5}{2}}$ 　　　　　　　　　8. $u(x) = \dfrac{128}{35\pi}x^{\frac{7}{2}}$

9. $u(x) = \dfrac{2\sqrt{x}}{\pi}\left(1 + \dfrac{8}{5}x^2\right)$ 　　　　10. $u(x) \approx \dfrac{2}{\pi}\sqrt{x}$

Exercises 6.3

1. $u(x) = x$ 　　2. $u(x) = x^3$
3. $u(x) = 6$ 　　4. $u(x) = \pi = 4x$
5. $u(x) = x$ 　　6. $u(x) = 4x + 14x^2$

Exercises 6.4

1. $u(x) = \sqrt{x}$ 　　　　　2. $u(x) = x^{\frac{3}{2}}$
3. $u(x) = \dfrac{1}{2}$ 　　　　　4. $u(x) = \sqrt{x}$
5. $u(x) = x^{\frac{5}{2}}$ 　　　　　6. $u(x) = x^3$
7. $u(x) = 1 + x$ 　　　　　8. $u(x) = 1$
9. $u(x) = x^2$ 　　　　　　10. $u(x) = \dfrac{2}{\pi}\sqrt{x} + \dfrac{15}{16}x^2$
11. $u(x) = 1 + x$ 　　　　12. $u(x) = x + x^3$
13. $u(x) = 1 + 3x^3$ 　　　14. $u(x) = 5 - x$

Exercises   6.5

1. $u(x) = x^2$
2. $u(x) = 10$
3. $u(x) = 10x$
4. $u(x) = 3 + 10x$
5. $u(x) = x + x^2$

Exercises   7.2.1

1. $u(x) = \dfrac{1 \pm \sqrt{1 - 2\lambda}}{\lambda}, \lambda \le \dfrac{1}{2};$
$\lambda = 0$ is a singular point, $\lambda = \frac{1}{2}$ is a bifurcation point

2. $u(x) = \dfrac{1 \pm \sqrt{-1 + 4\lambda}}{2\lambda}, \lambda \ge -\dfrac{1}{4};$
$\lambda = 0$ is a singular point, $\lambda = \frac{-1}{4}$ is a bifurcation point

3. $u(x) = \dfrac{1 \pm \sqrt{1 - 2\lambda}}{\lambda}, \lambda \le \dfrac{1}{2};$
$\lambda = 0$ is a singular point, $\lambda = \frac{1}{2}$ is a bifurcation point

4. $u(x) = \dfrac{3 \pm \sqrt{9 - 12\lambda}}{2\lambda}, \lambda \le \dfrac{3}{4};$
$\lambda = 0$ is a singular point, $\lambda = \frac{3}{4}$ is a bifurcation point

5. $u(x) = \dfrac{2 \pm 2\sqrt{1 - \lambda}}{\lambda}, \lambda \le 1;$
$\lambda = 0$ is a singular point, $\lambda = \frac{1}{2}$ is a bifurcation point

6. $u(x) = 2$
7. $u(x) = \sin x$
8. $u(x) = \cos x$
9. $u(x) = x, x + \dfrac{8}{3}$
10. $u(x) = x^2, x^2 + \dfrac{4}{3}$
11. $u(x) = x, x - 1$
12. $u(x) = x, x + \dfrac{1}{3}$

Exercises   7.2.2

1. $u(x) = 1 + \dfrac{\lambda}{2} + \dfrac{\lambda^2}{2} + \dfrac{5\lambda^3}{8} + \cdots$
2. $u(x) = 1 + \dfrac{\lambda}{4} + \dfrac{\lambda^2}{8} + \cdots$
3. $u(x) = 2\sin x$
4. $u(x) = 2\cos x$
5. $u(x) = \sec x$
6. $u(x) \approx 2x$
7. $u(x) = x^2$
8. $u(x) = x$
9. $u(x) = x$
10. $u(x) = x$
11. $u(x) = \sin x + \cos x$
12. $u(x) = \sinh x$
13. $u(x) = \cos x$
14. $u(x) = \sec x$

Exercises   7.2.3

1. $u(x) = x, 3x$

2. $u(x) = x^2, x^2 + \dfrac{12}{5}x$

3. $u(x) = x^3, x^3 + \dfrac{8}{3}x$

4. $u(x) = x, x + \dfrac{14}{3}x^2$

5. $u(x) = x + x^2, x + \dfrac{11}{3}x^2$

Exercises   7.3

1. $u(x) = 7\sqrt{\dfrac{x}{20}}, x + x^2$

2. $u(x) = \sqrt[3]{\dfrac{x^2}{24}}$

3. $u(x) = \sqrt[3]{\dfrac{4e^{2x}}{1 + e^{-4}}}, e^{2x}$

4. $u(x) = \sqrt[4]{\dfrac{3e^x}{1 - e^{-2}}}, e^x$

5. $u(x) = x, \ln x$

6. $u(x) = \dfrac{\sqrt{2}}{5}x, x \ln x$

Exercises   7.4

1. $u(x) = x$

2. $u(x) = x$

3. $u(x) = 1 - x$

4. $u(x) = 1 + x$

5. $u(x) = \sqrt[4]{\cos x}$

6. $u(x) = \sqrt{\sin x}$

Exercises   8.2.1

1. $u(x) = x^2$

2. $u(x) = x^2$

3. $u(x) = 1 + x$

4. $u(x) = 1 + x^2$

5. $u(x) = x^2$

6. $u(x) = e^{-x}$

7. $u(x) = 1 + x$

8. $u(x) = \sin x$

9. $u(x) = \cos x$

10. $u(x) = e^x$

Exercises   8.2.2

1. $u(x) = 3x$

2. $u(x) = 2x$

3. $u(x) = \sin x$

4. $u(x) = x^2$

5. $u(x) = x + \dfrac{1}{20}x^5 + \dfrac{1}{720}x^9 + \cdots$

6. $u(x) = 1 + \dfrac{1}{3}x^3 + \dfrac{1}{90}x^6 + \cdots$

7. $u(x) = 1 + \dfrac{1}{3}x^3 + \dfrac{1}{60}x^6 + \cdots$

8. $u(x) = x + \dfrac{1}{6}x^5 + \dfrac{1}{756}x^9 + \cdots$

9. $u(x) = 1 + 2x + \dfrac{5}{2}x^2 + \dfrac{1}{6}x^4 + \cdots$

10. $u(x) = 1 + x + x^2 + \dfrac{2}{3}x^3 + \dfrac{1}{6}x^4 + \cdots$

11. $u(x) = \sec x$ 12. $u(x) = \tan x$

Exercises   8.2.3

1. $u(x) = x$,   2. $u(x) = x^2$,   3. $u(x) = x + x^2$,   4. $u(x) = 1 + x$,
5. $u(x) = 1 + x$

Exercises   8.3.1

1. $u(x) = \pm \sin x$ 2. $u(x) = \pm e^x$
3. $u(x) = e^x$ 4. $u(x) = \pm(\cos x - \sin x)$

Exercises   8.3.2

1. $u(x) = \pm x^2$ 2. $u(x) = \pm x^4$
3. $u(x) = x^4$ 4. $u(x) = \pm e^x$
5. $u(x) = \pm x e^x$ 6. $u(x) = \pm e^{-x}$

Exercises   8.4

1. $u(x) = \pm x^3$ 2. $u(x) = x^4$
3. $u(x) = \sqrt[5]{1 + x^4}$ 4. $u(x) = \sqrt[4]{1 + x + x^3}$
5. $u(x) = \sqrt[5]{\cos x}$ 6. $u(x) = \sqrt[4]{\sin x}$

Bibliography

[1] G. Adomian, Solving Frontier Problems of Physics: The decomposition method, Kluwer, (1994).

[2] G. Adomian, Nonlinear Stochastic Operator Equations, Academic Press, San Diego, CA (1986).

[3] G. Adomian,and R. Rach, Noise terms in decomposition series solution, Computers Math. Appl., Vol.24,(11), (1992) 61-64.

[4] A. Alpinah and M. Dehghan, Numerical solution of the nonlinear Fredholm integral equations by positive definite functions, Appl. Math. Comput., 190 (2007) 1754–1761.

[5] J. P. Boyd, Chebyshev polynomial expansions for simultaneous approximation of two branches of a function with application to the one-dimensional Bratu equation. Appl. Math. Comput. 14, (2003) 189-200.

[6] Y. Cherruault, and V. Seng, The resolution of non-linear integral equations of the first kind using the decomposition method of Adomian, Kybernetes, 26 (1997) 109–206.

[7] Y. Cherruault, and G. Adomian, Decomposition Methods: A new proof of convergence, Mathl. Comput. Modelling 18(12), (1993), 103–106.

[8] S. Christiansen, Numerical solution of an integral equation with a logarithmic kernel, BIT, 11 (1971) 276–287.

[9] S. N. Curle, Solution of an integral equation of Lighthill, Proc. R. Soc. Lond. A, 364, (1978) 435–441.

[10] H. T. Davis, Introduction to Nonlinear Differential and Integral Equations, Dover, Publications, New York (1962).

[11] L. M. Delves, and J. Walsh, Numerical Solution of Integral Equations, Oxford University Press, London (1974).

[12] R. Estrada and R. Kanwal, Singular Integral Equations, Birkhauser, Berlin, (2000).

[13] J. Hadamard, Lectures on Cauchy's Problem in Linear Partial Differential equations, Yale University Press, New Haven, (1923).

[14] J. H. He, Some Asymptotic Methods for Strongly Nonlinear Equations, International Journal of Modern Physics B, Vol. 20, No. 10 (2006) 1141–1199.

[15] J. H. He, Variational iteration method for autonomous ordinary differential systems, Appl. Math. Comput., 114(2/3) (2000) 115–123.

[16] J. H. He, A variational iteration approach to nonlinear problems and its applications, Mech. Applic., 20(1) (1998) 30–31.

[17] J. H. He, A generalized variational principle in micromorphic thermo elasticity, Mech. Res. Com., 32(1) (2005) 93–98.

[18] J. H. He and X. H. Wu, Variational iteration method: New developments and applications, Comput. Math. Applic., 54 (2007) 881–894.

[19] A. Jerri, Introduction to Integral Equations with Applications, Wiley, New York, (1999).

[20] R. P. Kanwal, Linear Integral Equations, Birkhauser, Boston, (1997).

[21] R. P. Kanwal and K. C. Liu, A Taylor expansion approach for solving integral equations, Int. J. Math. Educ. Sci. Technol., 20(3) (1989) 411–414.

[22] S. Kobayashi, T. Matsukuma, S. Nagai and K. Umeda, Some coefficients of the TFD function, J. Phys. Soc. Japan, 10, (1955) 759–765.

[23] R. Kress, Linear Integral Equations, Springer, Berlin, (1999).

[24] J. M. Lighthill, Contributions to the theory of the heat transfer through a laminary boundary layer, Proc. Roy. Soc. Lond. 202(A) (1950) 359–377.

[25] P. Linz, A simple approximation method for solving Volterra integro-differential equations of the first kind, J. Inst. Maths. Applics., 14 (1974) 211–215.

[26] P. Linz, Analytical and Numerical Methods for Volterra Equations, SIAM, Philadelphia, (1985).

[27] K. Maleknejad and Y. Mahmoudi, Taylor polynomial solution of high-order nonlinear Volterra-Fredholm integro-differential equations, 145 (2003) 641–653.

[28] M. Masujima, Applied Mathematical Methods in Theoretical Physics, Wiley-VCH, Weinheim, (2005).

[29] G. Micula and P. Pavel, Differential and Integral Equations through Practical Problems and Exercises, Kluwer, (1992).

[30] R. K. Miller, Nonlinear Volterra Integral Equations, W. A. Benjamin, Menlo Park, CA, (1967).

[31] B. L. Moiseiwitsch, Integral Equations, Longman, London and New York, (1977).

[32] W. E. Olmstead and R. A. Handelsman, Asymptotic solution to a class of nonlinear Volterra integral equations, II, SIAM J. Appl. Math., 30(1) (1976) 180–189.

[33] D. L. Phillips, A technique for the numerical solution of certain integral equations of the first kind, J. Ass. Comput. Mach, 9 (1962) 84–96.

[34] D. Porter and D.S. Stirling, Integral equations: a practical treatment from spectral theory to applications, Cambridge, (2004).

[35] F. M. Scudo, Vito Volterra and theoretical ecology, *Theoret. Population Biol.*, 2:1–23 (1971).

[36] V. Singh, R. Pandey and O. Singh, New stable numerical solutions of singular integral equations of Abel type by using normalized Bernstein polynomials, Appl. Math. Sciences, 3 (2009), 241–255.

[37] R. D. Small, Population growth in a closed model, *Mathematical Modelling: Classroom Notes in Applied Mathematics*, SIAM, Philadelphia, (1989).

[38] K. G. TeBeest, Numerical and analytical solutions of Volterra's population model, SIAM Review, 39(3):484-493 (1997).

[39] H. J. Te Riele, Collocation methods for weakly singular second-kind Volterra integral equations with non-smooth solution, IMA Journal of Numerical Analysis, (1982), 437–449.

[40] A. N. Tikhonov, On the solution of incorrectly posed problem and the method of regularization, Soviet Math, 4, (1963) 1035–1038.

[41] H. Unz, Schlonilch's integral equation for oblique, incidence, J.Atmos. Terres. Physics, 28 (1966) 315–316.

[42] V. Volterra, Theory of Functionals of Integral and Integro-Differential Equations, Dover, New York, (1959).

[43] A. M. Wazwaz, Partial Differential Equations and Solitary Waves Theory, HEP and Springer, Peking and Berlin, (2009).

[44] A. M. Wazwaz, Linear and Nonlinear Integral Equations: Methods and Applications, HEP and Springer, Peking and Berlin, (2011).

[45] A. M. Wazwaz, A First Course in Integral Equations, World Scientific, (1997).

[46] A. M. Wazwaz, A reliable treatment for mixed Volterra-Fredholm integral equations, Appl. Math. Comput., 127 (2002) 405–414.

[47] A.M. Wazwaz, The decomposition method for approximate solution of the Goursat problem, Journal of Applied Mathematics and Computation, 1995, 299–311.

[48] A. M. Wazwaz, The variational iteration method for analytic treatment for linear and nonlinear ODEs, Appl. Math. Comput., 212 (1) (2009) 120–134.

[49] A. M. Wazwaz, The variational iteration method for solving linear and nonlinear systems of PDEs, Computers and Mathematics with Applications, 54 (2007) 895–902.

[50] A. M. Wazwaz, The variational iteration method; a reliable tool for solving linear and nonlinear wave equations, Computers and Mathematics with Applications, 54 (2007) 926–932.

[51] A. M. Wazwaz, The variational iteration method; a powerful scheme for handling linear and nonlinear diffusion equations, Computers and Mathematics with Applications, 54 (2007) 933–939.

[52] A. M. Wazwaz, The variational iteration method for a reliable treatment of the linear and the nonlinear Goursat problem, Applied Mathematics and Computation, 193 (2007) 455–462.

[53] A. M. Wazwaz, The variational iteration method for exact solutions of Laplace equation, Physics Letters A, 363 (2007) 260–262.

[54] A. M. Wazwaz, The variational iteration method for solving two forms of Blasius equation on a half-infinite domain, Applied Mathematics and Computation, 188(1) (2007) 485–491.

[55] A. M. Wazwaz, The variational iteration method for rational solutions for KdV, K(2,2), Burgers, and cubic Boussinesq equations, Journal of Computational and Applied Mathematics, 207 (2007) 18–23.

[56] A. M. Wazwaz, A comparison between the variational iteration method and Adomian decomposition method, Journal of Computational and Applied Mathematics, 207 (2007), 129–136.

[57] A.M. Wazwaz, The modified decomposition method for analytic treatment of non-linear integral equations and systems of non-linear integral equations, International Journal of Computer Mathematics, 82(9) (2005) 1107–1115.

[58] A.M. Wazwaz, Analytical approximations and Pade' approximants for Volterra's population model, Applied Mathematics and Computation, 100 (1999), 13–25.

[59] A.M. Wazwaz, A reliable modification of the Adomian decomposition method, Applied Mathematics and Computation, 102 (1999), 77–86.

[60] A.M. Wazwaz, The modified decomposition method and the Pade' approximants for solving Thomas-Fermi equation, Mathematics and Computation, 105 (1999), 11–19.

[61] A.M. Wazwaz, A new technique for calculating Adomian polynomials for nonlinear polynomials, Applied Mathematics and Computation, 111(1)(2000), 33–51.

[62] A. M. Wazwaz, Necessary conditions for the appearance of noise terms in decomposition solution series, Applied Mathematics and Computation, 81, (1997), 265–274.

[63] A. M. Wazwaz, The modified decomposition method and the Pade' approximants for solving Thomas-Fermi equation, Applied Mathematics and Computation, 105 (1999), 11–19.

[64] A. M. Wazwaz, The combined Laplace transform-Adomian decomposition method for handling nonlinear Volterra integro-differential equations, Applied Mathematics and Computation, 216 (2010) 1304–1309.

[65] A. M. Wazwaz, R. Rach, and J.-S. Duan, Adomian decomposition method for solving the Volterra integral form of the Lane–Emden equations with initial values and boundary conditions, Applied Mathematics and Computation, 219 (2013) 5004–5019.

[66] A. M. Wazwaz, R. Each, and J.-S. Duan, A study on the systems of the Volterra integral forms of the Lane–Emden equations by the Adomian decomposition method, Mathematical Methods in the Applied Sciences, 37(1) (2014) 10–19.

[67] A. M. Wazwaz, R. Each, and J.-S. Duan, The modified Adomian decomposition method and the noise terms phenomenon for solving nonlinear weakly–singular Volterra and Fredholm integral equations, Central European Journal of Engineering, 3(4) (2013) 669–678.

[68] A. M. Wazwaz, The Volterra integro-differential forms of the singular Flierl-Petviashvili and the Lane-Emden equations with boundary conditions, Romanian Journal of Physics, 58 (7/8) (2013) 685–693.

[69] A. M. Wazwaz, The variational iteration method for solving the Volterra integro-differential forms of the Lane-Emden equations of the first and the second kind, Journal of Mathematical Chemistry, 52 (2014) 613–626.

Index

Printed in the United States
By Bookmasters

Printed in the United States
By Bookmasters